Mysteries Beneath the Sea

Mysteries

BENEATH

THE SEA

William R. Corliss

THOMAS Y. CROWELL COMPANY
ESTABLISHED 1834
New York

Illustration on page 150 and quotation on page 151, from *In the Wake of the Sea Serpents* by Bernard Heuvelmans. *Le Grand Serpent-de-Mer* © Librarie Plon 1965. This translation © Rupert Hart-Davis 1968. Reprinted by permission of Hill and Wang, Inc.

Designed by Nancy Dale Muldoon
Manufactured in the United States of America
L. C. Card 71-127608
ISBN 0-690-57082-1
1 2 3 4 5 6 7 8 9 10

CONTENTS

CONTENTS

I

A NEW LOOK
AT THE EARTH

For centuries geology has been mainly a dry-land science. Rock strata deposited long ago beneath shallow seas and later elevated above sea level have been the source of most of our knowledge about the earth's past. In the 1950's and 1960's, however, more and more scientists—biologists, geophysicists, and archaeologists, as well as geologists—turned their attention to the sea. The results have been exciting; perhaps revolutionary is more accurate. Data from beneath the sea are changing our whole concept of how the surface of the earth is constructed and how it evolved down through the aeons. Since man is an earth creature, except for brief forays to the moon, these same data may also modify our thoughts about human evolution.

Most of the new data have had to be won laboriously from the bottom of the sea. Only with modern oceanographic research ships and new instruments have scientists really been able to come to grips with the sea bottom. They did not find the ancient, featureless sea floors they had expected. For example:

Fathometer surveys show that the earth's crust is cracked extensively beneath the sea. Further, deep trenches fringe volcanically active coastlines, and truly colossal submarine canyons have been cut in the edges of the continental shelves by some unidentified erosive agent.

Samples dredged from the deep-sea floor reveal very little sediment.

1

Radioactive dating of retrieved rock samples show that the present sea floors are only a couple hundred million years old—essentially brand new as geologists reckon time.

Magnetometers disclose that the earth's magnetic polarity has reversed frequently during the past several million years.

Paleontological studies of rocks deposited under ancient seas highlight frequent crises in the history of life, during which species were extinguished on a wholesale basis only to be followed by great forward spurts in evolution.

Such data combined with new archaeological findings above and below the sea and with the information contained in the moon rocks brought back by the astronauts have stimulated scientists to speculate that perhaps the earth's past has been more turbulent than hitherto supposed.

The earth is no longer viewed as a sphere with a solidified crust with continents and ocean basins frozen into a permanent pattern. Rather, the earth's surface is still plastic and still evolving under the influence of internally and externally generated forces. Life on earth is probably even more plastic than the planet's countenance, and it is doubtless responding to the same array of forces.

The possibility of external—that is, astronomically derived— forces affecting the evolution of the earth's physical features and its cargo of life creates the possibility that testimonial data may be useful in working out the very recent history of the earth. It is entirely possible that ancient man actually witnessed widespread physical upheavals, such as sea incursions and the impacts of large meteors, and recorded his observations in legend. This kind of information is usually studiously disregarded in scientific appraisals. It is fascinating and pertinent here to reexamine this kind of information in the light of the revisions being made to our planet's history by virtue of the bonanza of new evidence found beneath the sea. Legends and myths cannot, of course, be given nearly the same weights as magnetometer readings and paleontological evidence; but in trying to piece together our past, we should not ignore persistent hints, regardless of the source. Discretion must be the watchword here.

Scientific controversy is found in every chapter in this book. This should not be surprising in a field where an active revolution is acknowledged. Indeed, controversy is the scientific way. Never have important new scientific concepts been introduced without long debate. Einstein's relativity and Darwin's evolution are now well entrenched, but fierce were the battles along the way. In the same way, theories about continental drift, submarine canyons, and the extinction of species, in short, all of the subjects discussed in this book, are now being debated vigorously.

Underlying all of these controversies is an argument as old as geology itself: the conflict between Uniformitarianism and Catastrophism. The scientific philosophies inherent in Uniformitarianism and Catastrophism greatly influence the ways in which different scientists view the same set of facts, particularly the facts dredged up from the sea floor. Therefore, some background on these two "attitudes" is in order.

The philosophies of Catastrophism and Uniformitarianism offer competing incompatible interpretations of our planet's past, present, and future. At the heart of the matter lies a choice between a fickle, almost malevolent universe and one that is orderly and predictable, a world where man can control his own destiny. The key elements of each philosophy are shown below.

Catastrophism involves a fatalistic, almost mystical belief in a turbulent past and an unpredictable future. It is related to Fundamentalist creeds in which man is punished for his misdeeds by terrestrial upheavals.

Uniformitarianism incorporates optimism about the continuity of the world and the predictability of the physical processes that sustain man. Uniformitarianism is generically related to humanism, which claims man has the power to predict and build without the fear of violent perturbations from above or below.

Basically, Catastrophists believe that without cataclysms there can be no God; while the Uniformitarians maintain that without uniformity there can be no science. There is some middle ground between these extremes, but not much.

Originally, all scientists were Catastrophists. In the early 1800's came regimented Uniformitarianism. Today we see a few scientists beginning to fidget within the confines of the Uniformitarian mold. The history of the warfare between the two contending philosophies is fascinating and also vital to the way in which we interpret facts from sea bottom and mountaintop. In other words the facts themselves are indecisive and may still be fabricated into several reasonably consistent "explanations" of our cosmos. It should be emphasized, though, that Uniformitarianism is far and away the philosophy most acceptable to science. "If there is any circumstance thoroughly established in geology, it is, that the crust of our globe has been subjected to a great and sudden revolution." So wrote the French scientist, Georges Cuvier, in his *Essay on the Theory of the Earth* toward the end of the eighteenth century. He was one of the last and certainly the most famous of the Catastrophists to speak out and be heard before the doctrine of Uniformitarianism settled down over the scientific world. Cuvier opposed the evolutionary philosophy of his older contemporary Jean Baptiste Lamarck. The view of the Scot James Hutton that the earth's surface was the result of slow changes taking place over long periods of time did not impress him. History was a succession of catastrophes, Cuvier said. The latest one occurred five or six thousand years ago as described in the Bible. The frozen carcasses of mammoths discovered in the northern lands were ample evidence of the great changes that swept across the globe during the Deluge. Cuvier had a brilliant style, social charm, and considerable influence throughout Europe. He sustained Catastrophism against the anti-Biblical forces of evolution and Uniformitarianism then just starting to gather strength.

The heyday of Catastrophism came during the 1820's and 1830's under the inspiration of the English geologist William Buckland, the chief architect of the Catastrophist philosophy. Buckland exploited and extended Cuvier's researches and "returned natural history to the explicit service of religious truth." Buckland, his fellow Catastrophists, and many amateur geologists unearthed abundant evidence all over England purporting to show the truth of Biblical chronology. An inspired teacher, Buckland, in top hat and academic robes, led his classes to mudbank and cavern. His greatest work bore the impressive title *Reliquiae Diluvianae, or, Observations on the Organic Remains Contained in Caves, Fissures, and Diluvial Gravel, and on Other Geological Phenomena, Attesting the Action of an Universal Deluge.* The title neatly summarized his philosophy. The book appeared late in 1823.

While Cuvier and Buckland captured much popular and scientific attention, Uniformitarianism was slowly growing under the cultivation of James Hutton and Charles Lyell, two geologists who tramped up and down Scotland studying the record of the rocks. These two men had essentially the same kinds of facts available to Cuvier and Buckland; yet they arrived at totally different conclu-

sions. Hutton supposed that his Scottish Highlands had risen millimeter by millimeter over millions of years rather than being forci-
bly thrust up in a geological wink. The Uniformitarians admitted
to localized earthquakes and volcanoes, but by and large the
earth's wrinkled crust was the product of slow ponderous forces
still at work and within the ken of man. No mystical and violent
intercession from without for Uniformitarianism.

James Hutton was an amateur geologist, but in the 1700's many
of the great contributors to science were amateurs. In fact, geology
did not really exist until Hutton came along. A physician by training, he never practiced medicine, preferring instead agriculture,
chemistry, and, of course, geology. Hutton's major work was his
Theory of the Earth, which appeared in 1785. The seeds of Uniformitarianism contained in *Theory of the Earth* did not, however,
germinate immediately, for they contradicted the Biblical interpretation of the planet's past. It should also be said, though, that Hutton's writings were rather dull and never really captured a wide
audience. Catastrophism was not really challenged until the much
more persuasive Charles Lyell arrived on the scene a generation
later.

Charles Lyell was educated as a lawyer, but like his fellow Scot
Hutton, he soon became fascinated by geology and devoted more
and more of his time to his "hobby." Lyell explored Britain and the
Continent and, again like Hutton, became convinced that the past
could be explained in terms of processes he saw operating with infinite slowness around Europe. When he discovered Hutton's work,
he became a confirmed Uniformitarian. To exert any influence in
the early 1800's, a scientist had to write a book and enlist disciples
to promulgate the faith. The title of Lyell's book, which was published in 1833, left no doubt as to his philosophical leanings: *Principles of Geology, Being an Attempt to Explain the Former
Changes of the Earth's Surface, by Reference to Causes Now in
Operation.* After all, he was competing with Buckland's popular
book and he had to have an imposing title, too. *Principles of Geology* was well written and a "best seller"; it was one of the most in-
fluential scientific books of the century.

Science could not exist in a universe that was the plaything of
the gods, according to Lyell, who began gnawing away at the
foundations of Catastrophism with such assertions as: "Never was
there a dogma more calculated to foster indolence, and blunt the

keen edge of curiosity, than this assumption of the discordance between the former and the existing causes of change." In essence,
Catastrophism stood in the way of a rational understanding of nature; it was mystical, reactionary, and opposed to beneficial
change.

Uniformitarianism has remained dominant in science to the present day. However, every twenty or thirty years some individual,
either by his force of intellect or his persuasiveness, has managed
to release a few well-aimed missiles in the direction of Uniformitarianism. Inevitably these dissenters have been classified as cranks
or pseudoscientists. In these days when the pillars of Uniformitarianism show signs of strain, a really gifted Catastrophist might
bring down the whole temple. Three would-be Sampsons have already made the attempt: Ignatius Donnelly, George McCready
Price, and Immanuel Velikovsky; a prairie genius, a crusader for
creation, and an interpreter of legends, in that order. The works of
these men have been emphatically rejected by science, but we
should understand the efforts of Donnelly, Price, and Velikovsky
because their indiscriminate methods and extravagant claims have
instilled in science an instinctive reaction against anything smack-
ing of Catastrophism.

Of the three, Ignatius Donnelly was by far the most colorful and
talented. Born in Philadelphia, in 1831, he lacked professional
training of any kind, but he had tremendous energy and the ability
to remember everything that funneled into his brain. Politically active, he was once the Populist candidate for vice president. He had
three passions outside politics. First, he was convinced that Francis Bacon wrote most or all of Shakespeare's plays. His books *The
Great Cryptogram* and *The Cipher in the Plays* are cornerstones of
the Francis Bacon promoters who, like The Flat Earth Society, are
still peddling their wares. A second hobby that Donnelly pursued
with his customary ferocious zeal was proving the reality of Atlantis, that sunken city of Plato. His third love was the lore of Catastrophism.

During the short Minnesota summer of 1882, Donnelly churned
out the 450-page book *Ragnarok: The Age of Fire and Gravel*, in
which he anticipated Immanuel Velikovsky's 1950 work, *Worlds in
Collision*. In fact, the research techniques of the two men were
much the same. Both read widely and collected all pertinent facts,
half-facts, legends, myths, and even falsehoods without discrimina-

tion. Each wove his own pattern of past catastrophe, usually with scant regard for the validity of the data or the physical soundness of their theories.

George McCready Price had none of Donnelly's flamboyance. Hardly the inspiring orator or the persuasive politician, he was a dedicated Seventh Day Adventist. His whole life was taken up with serving his creed and undermining the perfidious doctrine of evolution. Price was raised by devout parents in Nova Scotia and began his geologic studies already certain that Genesis was the keystone of geology. Like Donnelly, Price was largely self-educated. Nevertheless, he was able to marshal impressive facts and (he thought) compelling logic against the hated doctrine of evolution. All geologic facts that filtered through his mind were turned toward one end: the destruction of evolution and the reinstatement of creationism. Price was the *authority* on geology that William Jennings Bryan cited during the famous Scopes trial in Tennessee. During the first half of the twentieth century George McCready Price was foremost among the antievolutionists.

The third of our Catastrophists, Immanuel Velikovsky, was born in Russia in 1895. With an M.D. from Moscow University, experience as a general practitioner in Palestine and later as a psychoanalyst, he brought to Catastrophism a tremendous capacity for library research but no field experience.

Velikovsky's basic message differs little from Donnelly's: the Flood and all of mankind's dim remembrances of past terrestrial convulsions stem from real celestial cataclysms. Both men culled the world's literature; Velikovsky spent some nine years at his task in the library at Columbia University. Employing great selectivity in choosing "facts" from mythology, he wove his tale of Catastrophism into his *Worlds in Collision*, published in 1950. *Worlds in Collision* was violently rejected by science.

Donnelly, Price, and Velikovsky are pertinent to the mysteries beneath the sea. As modern instruments have laid bare the ocean floor, the submarine canyons, the world-circling rift, the spreading sea floors, and other hints of past Catastrophism have been revealed. It may be that all we see under the sea *can* be explained by Uniformitarianism, but the wise might take a few side bets on some limited version of Catastrophism. The muted revival of George Darwin's theory that the Pacific basin is a scar left by the departing moon is a trend in this direction. Catastrophic mechanisms are no longer rejected out of hand. But science's built-in skepticism, its system of referred journals, and the collective weight of scientific opinion force the purveyor of a new theory to prepare a good argument.

READING LIST

CLARK, H. W.: *Crusader for Creation*, Pacific Press Publishing Association, Mountain View, Calif., 1966. (A biography of George McCready Price.)

GARDENER, M.: *Fads and Fallacies in the Name of Science*, Dover Publications, New York, 1957.

GILLISPIE, C. C.: *Genesis and Geology*, Harper & Brothers, Publishers, New York, 1959.

GRAZIA, A., DE, ed.: *The Velikovsky Affair: The Warfare of Science and Scientism*, University Books, Inc., New Hyde Park, N.Y., 1966.

LARRABEE, E.: "The Day the Sun Stood Still," *Harper's Magazine*, Jan. 1950.

NEWELL, N. D.: "Crises in the History of Life," *Scientific American*, Feb. 1963.

PRICE, G. M.: *Evolutionary Geology and the New Catastrophism*, Pacific Press Publishing Association, Mountain View, Calif., 1926.

RIDGE, M.: *Ignatius Donnelly; the Portrait of a Politician*, University of Chicago Press, Chicago, 1962.

VELIKOVSKY, I.: *Worlds in Collision*, Delta Books, New York, 1965, Pocket Book reprint.

2

THE CREATION OF THE
GREAT OCEAN BASINS

The earth's globe seen in schoolroom and library is more than two-thirds blue; no other planet circles the sun with such a watery countenance. In proportion, though, a few drops of dew on a September apple give it as much water as the earth. To ancient man venturing forth on the sea's fringes in his diminutive ships, this thin film was fathomless, dangerous, and full of mystery. To the modern scientist, trying to be more objective, the restless layer of salt water is 140 million square miles in area and averages two miles in depth. In exploring this liquid realm, man's ships have plied the surface waters for millenia and, within the past few years, have descended into the deepest ocean trenches. Only now are scientists beginning to realize how many mysteries this thin veil has concealed. Beneath the surface are clues to the earth's origin, the beginning of life, and even man's place in the whole drama. The best place to begin the tale is with the oceans themselves.

If you hold a globe upright with the North Pole pointing upward and spin it slowly, you will see that the northern half is a land hemisphere and the southern half is largely water. In fact, roughly five-sixths of the earth's land mass is antipodal to or diametrically opposite ocean rather than land. C. G. A. Harrison, in the September 9, 1966, issue of *Science*, computed that there is only one chance in fourteen that this odd distribution of land and water is a random occurrence. When dice do not fall according to statistical laws, they may be loaded. Here, too, there may be some special physical reason for the odd distribution of land and water.

Giving the globe another twirl, strangely triangular continents swing past the observer's view. In 1899, the British geologist, John W. Gregory, described the ocean-continent arrangement as "a pair of cogwheels with interlocking teeth." Indeed, some of the continents look as though they might fit together like jigsaw puzzle pieces, particularly the Americas, Europe, and Africa. Imagine being in a satellite and looking down on the immensity of the Pacific from a stationary orbit 22,000 miles up; you would see only blue water. It would seem as if a whole side of the earth had been blasted out or torn away and invaded by the sea. The rim of the Pacific basin is roughly circular, except for invading Australia. The Pacific is some 7,000 miles across and surrounded by volcanic and earthquake belts—a "ring of fire" as the great geologist Eduard Suess called it. By contrast the roughly parallel sides of the "Atlantic river" are volcanically quiet.

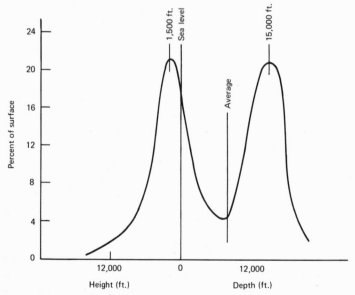

The earth possesses a marked two-tiered surface structure. The theories of continent and ocean-basin formation attempt to explain this feature.

In addition to such perplexing gross features, careful measurements of the ocean depths and the continents show distinctly that the earth is a two-tiered planet; with much of the land at fifteen hundred feet *above* sea level and the sea bottom at fifteen hundred feet *below*. Seismologists also know that the rock foundation under the oceans is heavier, thinner, and of different composition and structure than that under the continents. This sampling of facts leaves no doubt that the ocean basins are distinctly different from the continents; so different that science has been trying to explain the disparities for centuries. The ocean basins are not just "uneven" spots on a rough sphere, like blemishes on our dewy apple, but singular structures calling for a special hypothesis that explains the differences. Through the years, many such hypotheses have been proposed; some good, some fanciful; but all intimately related to the history of the earth and the evolution of life upon it.

DIMENSIONS OF THE OCEANS AND SEAS °

BODY OF WATER	AREA (MILLIONS OF MILES ^{2})	VOLUME (MILLIONS OF MILES ^{3})	MEAN DEPTH (FEET)
Atlantic Ocean °°	41.2	85.2	10,860
Pacific Ocean °°	69.5	173.5	13,200
Indian Ocean °° '	28.9	70.2	12,760
Arctic Ocean	5.5	4.1	3,960
All adjacent seas plus Arctic Ocean	15.4	11.6	3,960

° Data extracted from *The Encyclopedia of Oceanography*, R. W. Fairbridge, ed., Reinhold Publishing Corp., New York, 1966.
°° Including adjacent seas and Arctic Ocean.

Before the Renaissance, when only the gross features of the earth were known, there seemed little need for any elaborate account of how the ocean basins were formed. According to Genesis, God created the earth and the waters thereon. The earth was not perfectly smooth and the waters naturally filled in the low places; just as the puddles in the fields after heavy rains. The Renaissance, however, brought forth inquiring minds that asked why, how, and when; these minds did not accept the orthodoxies of the past. In the seventeenth century natural scientists, such as René Descartes and Gottfried Leibnitz, tried to explain the ocean basins in terms of the processes and phenomena they observed in the world about them.

Robert Hooke, who was a fertile source of theories about the earth and the moon, which brought him into violent collision with Newton, suggested that earthquakes had transformed ancient continental areas into sea bottoms; that is, immense blocks of the earth's crust had foundered and been covered by water. Leibnitz had a different idea: the earth's crust was created with huge subsurface bubbles of air and water; when these collapsed, the oceans were born. Descartes' theory possessed ingredients that reappeared some 350 years later in the twentieth century. He postulated that during the early history of the earth the sun's heat was so potent that it caused the earth to expand, cracking the crust, forming widening gaps that filled with water.

These seem naïve hypotheses to us today. How could they be anything else when the Americas were little more than phantom land masses to the far west and the Pacific was the domain of Polynesian outriggers? The first sophisticated theories had to wait for more data.

In the seventeenth and eighteenth centuries European explorers sailed into every sea and sketched in the land-ocean frontiers missing on world maps. But the great voyages of James Cook and the other sea explorers provided science with only surface maps and spare soundings of the sea bottom. It was not until the 1870's that the deeps began to be sounded in earnest. Within that decade the English *Challenger*, the German *Gazelle*, the American *Tuscarora*, and other oceanographic ships brought back the first comprehensive surveys of the sea bottom, the sea itself, and the life it contained. This was the kind of information that science needed to reconstruct the history of the ocean basins on a more objective level.

Before tracing the development of some of the more important theories, the scientific stage must be set. For background music the discipline of geology provided the guiding principle of *Uniformitarianism*. This principle stated that existing geologic processes, as observed in the field, acting with almost infinite slowness, were sufficient to account for all geologic stages. The Catastrophists, on the other hand, interpreted the Bible literally when it told of vast upheavals in the earth's past. All about them they saw grotesquely contorted strata and myriads of fossils that told them of past terrestrial convulsions, thus confirming divine revelation. But the Uniformitarians told them that they had read the rocks wrongly. Here the important point is that today's catholic geologists and geophysicists are Uniformitarians. Any theory of ocean-basin formation involving Catastrophic meteors or large-scale rending of the earth's crust must be considered unorthodox.

Allied in a strange marriage with Uniformitarianism were the Catastrophic "molten earth" theories that depended upon a passing star or solar indigestion to pull out or eject the raw materials for the planets from the sun. The American scientists Thomas C. Chamberlin and Forest R. Moulton proposed the solar ejection theory in 1900. It replaced the long-favored Nebular Hypothesis formulated by Immanuel Kant and Pierre Simon de Laplace in the late 1700's. Kant and Laplace had envisaged a hot primordial earth formed when a nebula condensed into the sun and its family of planets. A hot, initially molten, earth would have been plastic. One could easily imagine it solidifying and the vapors around it condensing into oceans. Uniformitarianism and the malleable virgin earth formed the foundation for many theories of ocean-basin formation.

In the 1940's and 1950's, however, the German astronomer Carl von Weizsäcker proposed a "cold-earth" theory, wherein the earth and other planets were formed by the *accretion* or gradual accumulation of dust particles. Harold C. Urey and others elaborated on the accretion hypothesis, and today it is accepted by most astronomers. In a sense, this hot-to-cold switch represented a return to pure Uniformitarianism because the formation of the earth was then something that happened slowly under the influence of normally prevailing forces. Planetary materialization became an event that could and probably did happen around most stars; no special astronomical cataclysm was required. A cold earth is not easy to mold into continents and ocean basins, though; and new ocean-basin theories or at least variations on the old, hot-earth hypotheses were in order.

First, though, the traditional molten-earth scenario:

Act I: The molten earth begins to cool and chemical and gravitational mechanisms begin to separate the iron core from the lighter materials destined for the crust.

Act II: Blocks of light continental material are differentiated from the heavier sea-bottom material; both float on a "sea of plastic magma."

Act III: Cooling continues and the lighter, higher floating continents and heavier sea floors are frozen into place.

Act IV: In the very long final act, beginning some hundreds of millions of years after the curtain was raised, the earth cools down to its present temperature (taking about 4.6 billion years). In doing so, its circumference contracts tens, perhaps hundreds of miles, pushing up mountains and wrinkling the ocean floor, and kneading the strata we see today.

Reviewer's Notes: This play has a great deal of appeal; it is essentially Uniformitarian; it is based on solid physical laws; it explains much the geologist sees in the field. Even Newton thought that mountains might have been born as the result of a cooling and contracting earth. Many of today's geologists vote with Newton, even in the face of considerable modern evidence for an expanding earth which will be introduced later.

The above portrait of a cooling, slowly solidifying earth provides us with a mechanism that separates the sea floors from the continental masses, but it says nothing about the peculiar shapes and unusual distribution of continents and ocean basins. Strangely, modern theorists have not worried as much about the gross geometric features of the earth as the generalists of the last century—that period when scientists were still called natural philosophers. To a natural philosopher explaining the triangular continents and their antipodal relationship to the ocean basins was an irresistible challenge.

One such natural philosopher, a Scotsman by the name William Lowthian Green, suggested in the 1857 volume of the *Edinburgh New Philosophical Journal* that the basic geometric form of the earth tended toward a pyramid. In his 1875 book, *Vestiges of the Molten Globe,* Green explained his tetrahedral earth more thoroughly. It would be natural, he said, for a cooling, contracting sphere with a solid crust and liquid interior to assume a geometric form that had the same volume as the sphere but a larger area. In other words, the crust would not wrinkle indiscriminately like a prune's as the core shrank, but it would "prefer" a simple geometry, just as crystals do or the strangely polygonal patterns of ground in the arctic. Green expected nature to be simple and symmetrical—a pandemic notion among most scientists, even today. At the apices of Green's tetrahedron were the land masses of

North America, Europe, Asia, and Antarctica; on the faces were the great four oceans (Atlantic, Pacific, Indian, and Arctic). There were extra pieces left over (South America and Australia), and the tetrahedron theory never gained much popularity, although dodecahedrons and other polyhedrons were also tried.

The search for natural order in the earth's topography continued when, in 1885, Stanley Grimes's book *Geonomy: Creation of the Continents by the Ocean Currents* was published by Lippincott, in Philadelphia.* Grimes's theory, called the "elliptical theory," harked back to a time when the earth had cooled. The crust had solidified and was covered entirely by water. The same natural forces that create the great cyclones and anticyclones in our atmosphere would, Grimes reasoned, establish circulation patterns in the global ocean. He envisaged six somewhat elliptical cells, three clockwise cells in the Northern Hemisphere, three counterclockwise in the Southern Hemisphere. The three pairs of currents scoured the rock beneath for aeons, and the erosion products collected in the centers of the six ellipses. The collected sediments pressed down upon the still plastic magma beneath. The magma, having nowhere else to go, pushed up on the earth's crust at the boundaries of the elliptical currents. The upwardly surging crustal segments broke through the ocean surface and eventually became the continents we see today. The Gulf Stream and the other great ocean currents are remnants of Grimes's elliptical currents.

There are many objections to this hypothesis; for example, it implies that continental crust and ocean-bottom crust are the same, and this is definitely not the case. And we know from seismic studies that the continents have deep solid roots extending well below the bottom of the ocean-floor crust. The continents are not balancing precariously upon a lens of liquid magma; rather, they float like ice cubes, immobilized to some degree by a thin layer of heavier ocean-bottom crusts. What is more important than the validity of Grimes's hypothesis is his determination to find a physical reason for the gross features of the earth's surface.

On this note of dynamism and change, the discussion moves easily toward two extreme classes of theories about ocean-basin for-

---

* "Theistic evolution" was another theoretical creation of Grimes. In 1850, eight years before Darwin belatedly published *The Origin of Species,* Grimes had proposed that the higher animals had evolved from the lower animals, although he maintained belief in God, contrary to many of the later evolutionists.

mation. The mechanisms proposed are dynamic; in fact, they fall in the realm of Catastrophism. The first class includes theories stating that matter was torn from the earth, leaving water-filled scars behind; the second class encompasses hypotheses wherein projectiles hit the earth, gouging out craters.

Catastrophism is still a bad word in the geologist's lexicon, but not as bad as it used to be. Catastrophism leaves little room for neat, orderly theories of the world. Back in 1925 geology was so calcified that William Morris Davis felt compelled to jolt his colleagues with a paper entitled *The Value of Outrageous Geological Hypotheses.* Unfortunately, it was not until the 1950's that oceanographers and geophysicists began to break the monotony by reviving thoughts about mobile continents and turbulent terrestrial viscera. The whole attitude of the earth scientists has changed markedly. For example, in 1961 the American geophysicist Robert S. Dietz said, "The ocean basins seem like gaping wounds in the earth's skin, exposing the raw, moving flesh of the earth." Hardly a stodgy description; one might say it definitely has Catastrophic overtones.

What could be more Catastrophic in our history than some vast wrench or convulsion that tore or threw off from the earth a glob of planet-stuff the size of the moon? This hypothesis seems outrageous and, on top of that, contrary to all the tenets of Uniformitarianism. Yet, the Pacific Ocean basin may have been the womb of the moon.

The thought that the earth gave birth to the moon is usually attributed to George H. Darwin, the son of Charles. Actually, the idea is much older. In 1857 an American named Richard Owen stated that the moon was thrown off the earth, leaving the Mediterranean basin behind as a scar. Darwin employed astronomical reasoning. Noting that the moon is slowly receding from the earth, he extrapolated backward in time, and inferred that the moon must have been much closer, perhaps even part of the earth. Darwin proposed, in 1879, that the material now forming the moon was actually thrown off the earth in the dim geologic past when the earth was molten and the solar-induced tide and the earth's period of rotation became equal (resonated). Three years later Darwin's idea was elaborated by the British geophysicist the Reverend Osmond Fisher with the publication of his paper "On the Physical Cause of the Ocean Basins," in the January 12, 1882, issue of *Nature.* The

Reverend Mr. Fisher had noticed that the moon's low density (3.3) was close to that of the earth's crust (2.5); the moon, therefore, might have been torn off after the light crust had solidified. Besides, he went on, the land masses on either side of the Atlantic looked as though they might have fractured and drifted apart following some cataclysm, such as the loss of a huge piece of crust from the region of the present Pacific Ocean. The Reverend Mr. Fisher also believed that continental fission was associated with the distribution of the species and the evolutionary pressures postulated by George Darwin's father.

The American astronomer, William H. Pickering, who was a prolific propounder of outrageous hypotheses, supported the moon-birth idea. In a 1907 paper Pickering claimed that the lack of sial (silicon-aluminum continental rock) in the Pacific basin was another consequence of the loss of the moon. Further, he believed that fully three-quarters of the earth's crust was torn away during the catastrophe; this fraction is very nearly the fractional area of today's ocean basins.

Another intriguing thought was injected by Oliver J. Lee in 1930 in the magazine *Science*. Lee stated, "A remarkable asymmetry exists in the longitude of the earth's magnetic poles, which are presently in 96° west and 155° east longitudes. They are, therefore, only 109° apart, and their longitudes mark out roughly the average boundaries of the Pacific Ocean." Lee went on to propose that enough deep-lying magnetic material was torn away from the earth to fix the magnetic poles in their peculiar positions. Here we see an early attempt to use the earth's magnetic fields as a gauge of the planet's past history.

No longer can the subject of amateurs, pseudoscientists, cranks, and real but maligned prophets be avoided. They lurk wherever Catastrophic theories of the earth are discussed. There has always been open warfare between established science and proposers of outrageous hypotheses; perhaps there has to be if the march of science is to be orderly and fruitful. Today's crank may be tomorrow's hero, but cranks and pseudoscientists are legion, and science has no choice but to force each purveyor of offbeat ideas to present good physical evidence supporting his pet theory. No ore-separation process is perfect, however, and occasionally some nuggets are lost. And, on occasion, science overreacts. Whatever the reason, some of these discarded tailings have great historical interest and some may yet turn out to be valuable. Because of their questionable pedigree, these tidbits are herewith properly labeled, much as we now label cigarette packages with cautionary notes.

The above diversion was necessary because it is time to introduce a contraband notion: that the moon may have been born when a celestial body brushed close enough to the earth to gravitationally pluck off a section of its crust. This is analogous to the old and now discounted tidal encounter explanation worked out by Chamberlin and Moulton to account for the origin of the sun's planets. The trouble is that our current view of solar-system history has no room for interlopers big enough to wrench the moon from the earth. More specifically the dogma states that the planets have been in their present orbits since the birth of the solar system. This hard line taken by the astronomers does not deter nonscientists from proposing interplanetary near misses.

In 1932, in a perfectly fascinating, privately produced, mimeographed book entitled *The Atlantic Rift and Its Meaning,* an amateur named Howard B. Baker suggested that the moon was born at the end of the Miocene period (some eleven million years ago), when a gravitational encounter occurred between the earth and Venus. The subject of Venus and its possible indiscretions in the astronomical past can hardly be discussed with equanimity with most physical scientists because of the famous Velikovsky incident of the early 1950's.

Velikovsky claimed in his book *Worlds in Collision* that ancient legends of terrestrial upheavals proved that Venus had sideswiped the earth in the past. Legends are hardly scientific data, and scientists violently rejected the notion of a wandering Venus. The fact of interest here is that Venus was tagged as a possible astronomical interloper long before Velikovsky came along. Some other solar system troublemakers will be introduced shortly.

Meanwhile, how does the moon-birth theory fare in the light of today's larger fund of facts about earth and moon? Very well, it seems, according to a detailed review made by John F. Simpson in the British magazine *Spaceflight.* Simpson begins by reviewing earth-moon dynamics, the recession of the moon, and the tidal-resonance theory. This century-old kind of analysis follows Darwin's footsteps. To this, Simpson adds magnetic data taken around the edge of the Pacific basin. The wrinklelike linear magnetic anomalies paralleling the coasts indicate (to him) encroaching continental land masses that are drifting in to fill up the hole left by the moon. He also mentions the asymmetry of the earth's field and Suess's "ring of fire" around the Pacific basin as supporting evidence. All in all, Simpson's review marshals an impressive amount of supporting evidence for the moon-birth theory. Simpson wrote this review in 1963 and could not know that the Apollo astronauts would return with lunar rocks with compositions similar to that of basalt, which supposedly underlies the ocean basins.

Is all this scientific proof? No; the best that can be said is that our best data do not rule out the moon-birth theory. Actually, the Apollo findings have given Darwin's moon hypothesis a credibility it has not had for decades. However, the great age of some lunar rocks (about 4.5 billion years) indicates that if the moon was torn from the earth, the cataclysm occurred very early in the earth's history.

There are other theories accounting for the presence of our outsized satellite. The theory currently accepted by most scientists states that the moon itself is an interloper and, in a close brush with the earth, was gravitationally captured. Supporting evidence here is not conclusive either. Another possibility is that the earth and moon were formed together through the gradual accretion of matter.

Great clouds of dust and rocky debris still wander through outer space. Each day, something like one thousand tons of this material drift down through the atmosphere to the earth's surface. Larger chunks occasionally flash through the atmosphere and arrive intact at the earth's surface. Some meteorites weigh tons; some are big enough to gouge out holes the size of Meteor Crater in Arizona. Other great craters dot the earth, though erosion has softened the features of most. Hudson Bay may be an ancient meteor crater. In addition, our celestial neighbors, the moon and Mars, have apparently been subjected to interplanetary bombardment that makes the earth's craters seem puny by comparison. The great dark "seas" or maria of the moon are roughly circular and are hundreds of miles in diameter. Pictures from the space probe, Mariner 4, show craters at least seventy-five miles in diameter on Mars; and, of course, Mars has its maria, too. Did the earth somehow escape the largest missiles?

This question introduces the most controversial of the ocean-basin hypotheses: that of meteor impact. The concept of an onslaught against the earth from space is even older than the Bible, wherein one finds ample warning of retribution via fire and brimstone from the skies. Scientists, too, have toyed with the possibility of celestial missiles. In 1694 Edmund Halley (of comet fame) read the following paper before the Royal Society: "Some Considerations About the Cause of the Universal Deluge." Halley believed that an encounter with a comet caused the Flood and perhaps moved the earth's axis at the same time. Laplace panicked Europeans with tales of tidal waves two and a half miles high caused by comets sideswiping the earth. These were the days when Catastrophism reigned supreme; comets and other celestial phenomena were readily accepted as causes for the terrestrial convulsions recorded in history, legend, and the contorted rocks seen throughout the world.

Modern science cannot deny the possibility—remote though it may be—of impacting comets and huge meteors. Planetoids such as Hermes, Adonis, and Icarus admittedly make astronomers edgy. In 1968 Icarus passed within 4 million miles of the earth. Hermes, which is just under a mile in size, zipped past at 475,000 miles in 1937. An impact with a billion-ton Hermes at a velocity of several tens of miles per second would make a crater many times the size of the projectile itself as well as trigger floods, earthquakes, and other terrestrial disasters.

Despite the past and present possibility of large meteors impacting the earth, scientists have generally shied away from associating them with the creation of ocean basins. Hudson Bay, the famed Carolina Bays, and other relatively small depressions in the earth's topography have been ascribed to meteors, but only a very few scientists believe that the ocean basins may have been blasted out in the same manner. Scientists, such as Robert S. Dietz and E. R. Harrison, have published a few speculative papers, but generally silence reigns.

Outside the pale of the established scientific community, Allan O. Kelly has published privately several booklets describing the effects of meteor collisions on the earth. In his 1953 book (written with Frank Dachille), *Target: Earth*, the ocean basins are treated as meteor craters. Generally, these publications in and out of the scientific literature have made little impression.

The placid, steadily flowing river of science was ruffled in 1961 when an article by John J. Gilvarry, a Ph.D. physicist, was published in the November 4 issue of the *Saturday Review*. The title of the article alone was enough to raise the hackles of many scientists: "How the Sky Drove the Land from the Bottom of the Sea." Without going into all the details, Gilvarry basically said: there are many big meteor craters on the earth, and the ocean basins should be counted among them. The meteor hypothesis, he said, is consistent with the character of the ocean floors, the ocean-basin shapes, the precipitous walls of the basins, and the fact that the sea-floor crust is much thinner than that of the continents. In addition, the overall dimensions of the ocean basins fall on the same depth-diameter curves as those of the lunar craters and the big holes created by nuclear weapons here on earth, a fact indicating a similar Catastrophic origin.

Gilvarry's hypothesis had been published in *Nature* several months before the *Saturday Review* article appeared. *Nature* often publishes speculative articles, and Gilvarry's suggestion evoked no

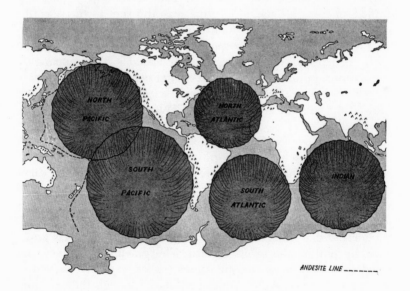

ANDESITE LINE _____

Gilvarry suggested that the earth's major ocean basins were blasted out by meteors. The Americas, however, look as if they might have fitted together with Europe and Africa at one time. The loss of crustal material in the Pacific region, due to meteor impact or the loss of the moon, might have caused the continents to drift apart. (Reproduced with permission from J. J. Gilvarry, *Saturday Review*, Nov. 4, 1961, p. 53; drawing by Doug Anderson)

untoward excitement. The *Saturday Review* article, however, stimulated seven well-known scientists (including Nobelist Harold C. Urey) at the University of California, at La Jolla, to write a letter of rejoinder and protest. The letter pointed out that the ocean basins (particularly the Atlantic) are not really very circular and that the large, extensive midocean ridge systems hardly square with what one would expect to find at the bottom of meteor craters. Why was the nicely circular Arctic Ocean ignored, they asked. Because it did not fit the depth-diameter curve? Further, the letter went on, Gilvarry had assumed water covered much of the moon, when the craters were blasted out, as it did on earth, but there was no evidence for large quantities of water on the moon in its past.*
The rather testy critique stated firmly that Gilvarry's "speculations are unacceptable to most if not all competent students of the origin and history of the earth."

The critical letter from La Jolla may have been technically justified—indeed, a detailed critique was also sent to the scientific journal *Nature*—but it also took *Saturday Review* to task for letting the general public see a speculative idea "without the aid and advice of specialists in the field discussed." This apparently smacked of suppression of free speech and censorship by the scientific establishment. *Saturday Review*'s science editor, John Lear, followed up with an article entitled, "When Is a New Idea Fit to Print?"

Since the above confrontation took place, the American Lunar Orbiter spacecraft have made gravimetric studies of the maria on the near side of the moon. The results show that the maria differ radically from terrestrial ocean basins. Terrestrial antennas tracked the Orbiters with great precision as they swung across the moon's face. Whenever their orbits dipped slightly toward the surface, it was inferred that some extra mass near the lunar surface was gravitationally pulling the orbits inward. The gravimetric surveys revealed that large mass concentrations (called *mascons*) existed at the center of every one of the ringed maria, though they are rare elsewhere on the moon. The obvious inference is that the large meteors that created the maria are now buried not far beneath the surface. For example, if the Mare Imbrium mascon is assumed to be thirty miles below the surface, it would be roughly sixty-two miles

---

* The lunar soil and rock specimens confirm that the maria have never been covered with water.

in diameter (assuming a nickel-iron composition). A meteor of this size is easily capable of blasting out a crater several hundred miles in diameter. Now, if the earth's ocean basins show similar evidence for buried masses, scientists would be much more inclined to accept the proposition that they too were created by the same mechanism, perhaps even at the same time, as the lunar craters. But gravimetric surveys of the earth's ocean basins show no evidence of buried objects at the basin centers. To the contrary, the Pacific basin shows a mass deficiency in the northeast and a mass excess near the southwest boundary. Of course, there is always the possibility that the terrestrial continents have drifted over the buried projectiles, concealing or at least distorting the evidence.

An interesting variation of the meteor theory has been proposed several times over the years: instead of falling masses excavating holes in the crust to form seas, perhaps the masses just plopped softly onto the crust and became the nuclei of our continents. An asteroid the size of Ceres (470 miles in diameter), for example, might fall on a plastic earth like a raisin on hot oatmeal to create Australia or Antarctica. Or possibly the moon was torn apart by terrestrial tidal forces in the past, depositing low-density pieces on the earth to become continents. This, of course, is almost the reverse of Darwin's theory for the origin of the Pacific. None of these extraterrestrial deposition theories are taken seriously today, though they add a fillip to the end of the remarkably diverse and contentious array of ocean-basin theories.

Large meteors undeniably hit the earth and leave scars on the crust, but the weight of the evidence presently at hand seems to deny that our ocean basins were created in this way—lunar craters a few hundred miles across, yes; but terrestrial basins a few thousand miles across, no. We must, however, recognize that meteors are one of the cornerstones of Catastrophism, which is again rearing its ugly head to the concern of all good Uniformitarians.

Fascinating though these Catastrophic theories are, an acceptable hypothesis of ocean-basin formation must be compatible with modern data. These data imply that the earth gradually accreted as cold bits of matter circling the sun were gravitationally drawn together. Manifestly, the earth is no longer a random conglomeration of interplanetary flotsam and jetsam; it has a definable structure, a core, and a crust with low-density continents and high-density sea floors. As the geophysicists say, the earth has "differentiated." Such structure seems possible only if the earth were once molten, so that lighter materials could float to the surface and chemical reactions could take place. The kinetic energies of the cold particles striking the proto-earth might have caused some melting, but this source of heat seems inadequate to melt the whole planet.

A better source of heat was discovered by Henri Becquerel in 1896: radioactivity. Worldwide surveys during the first half of this century showed that the earth's complement of radioactive isotopes, particularly potassium-40, which has a half-life of 1.3 billion years, may have been sufficient in the past to liquefy the earth. Heat-flow measurements on land and sea bottom indicate this outwelling of heat continues and may be increasing.

If the earth is heating up rather than cooling down, it must be expanding rather than contracting; and we have a new philosophical ball game. Even before the recent discoveries of great cracks in the sea bottoms and their apparent spreading tendencies, there was ample dry-land evidence that tension forces exist in the earth's crust. As early as 1927 J. Barrell pointed out that the great African Rift Valleys, the Mozambique Channel, Death Valley, the Davis Strait, and other cracks in our planet's crust signified expansion of the earth. The trouble was the worldwide mountain chains and the Pacific island arcs and their associated deep-sea trenches seemed to testify that, to the contrary, great compressional forces had caused the crust to buckle and wrinkle like a rug on furniture-moving day. How could both tension and compression exist at the same time?

The Irish geologist John Joly was one of the first to draw a theoretical model of the earth employing radioactive heating. In his 1925 book, *The Surface History of the Earth*, Joly proposed a scenario like this:

Prologue: Radioactive heat has liquified the earth. The continents and ocean floors have been separated by chemical and gravitational differentiation. The earth has cooled and the continental masses are frozen in position by the solid basalt below.

Act I: The radioactively generated heat no longer escapes easily by liquid convection and must depend upon thermal conduction to take it through the solid surface of the earth, where it is radiated to cold outer space. Thermal conduction through the solid crust is less efficient in transporting the heat, and internal temperatures begin to rise again.

Act II: The basalt sea upon which the continents perch partially liquifies. With the increase in temperature, the earth expands, causing cracks to develop. The basalt's density drops and the floating continents sink. The seas encroach upon the continents, depositing the marine rocks and fossils so evident the world over. The continents may also "drift" on their basaltic lubricant, but Joly did not make continental drift an integral part of his theory. (The theory of continental drift was in disfavor at the time.)

Act III: With the partial liquification of the basalt, the radioactive heat again escapes easily to the surface, where it dissipates into space. The earth once more cools and solidifies.

Act IV: As the basalt sea cools, it contracts, and the continental crust buckles and folds as it subsides into the empty spaces left by the shrinking basalt. Is Act IV the Finis?

Hardly—Joly's scheme is cyclic, and the play repeats as long as there is enough radioactive material left to stoke the furnace. This cyclic melting and solidifying squares nicely with the geologists' belief that the world has undergone many "revolutions" or cycles of mountain building. Joly's heat-powered machine, which reminds one of the mechanism driving Old Faithful and other periodic geysers, is one of the few that give the earth a heartbeat—something steady and internal that causes periodic external cataclysms. The theory appeals to both the Uniformitarian and the Catastrophist.

Few geologists and geophysicists speak of Joly's theory any more; they do not have sufficient data to prove or disprove it. Instead, interest has focused on models of the earth that combine the circulation cells of Grimes and the radioactive heat of Joly, though these theoretical precursors are rarely given their due.

A key ingredient in most modern models of ocean-basin and continent formation is the convection cell. When a pan of water is heated on the stove, the hotter and less dense fluid rises toward the surface, while heavier and cooler surface water sinks toward the heat source. If conditions are right, a geometrical pattern of rising and descending currents will be established; each geometrical element being called a cell. It is possible that similar convection cells exist under the earth's solid crust. The earth's stove, of course, is the radioactive potassium distributed throughout the entire planet.

In one favored model of the earth, ponderous currents of viscous, barely plastic magma rise under some parts of the solid crust, liberating their loads of radioactive heat as they turn and creep horizontally for thousands of miles beneath the earth's solid surface. The cooled, heavier magma finally sinks into the depths again where it will pick up more heat and repeat the cycle. Current belief is that upwellings occur beneath the ocean floors, where high heat flows are measured in the crust, and that magma descends under the continents. The magma may only flow a few inches each year, but the creeping currents may carry continental land masses like conveyor belts and stretch the sea floors in areas where they emerge from below. Here, it is the earth's internal machine that is important. It is interesting to note in passing that the earth's motor may also be the electrical generator that creates the magnetic field surrounding the earth. The more different things a hypothesis explains, the better it is.

Compare the convection cells of viscous magma to the oceanic circulation cells proposed by Grimes nearly a century ago. Instead of being built from water-carried sediment a la Grimes, the continents might have been deposited little by little over the aeons around nuclei located where the cold magma descends back into the earth. Think of the continents accreting age after age from lighter sialic (silicon-aluminum) minerals brought up from the depths of the earth. The continents would then be like pearls, old in the center and young outside. Studies of North America by A. E. J. Engel, a geologist at the University of California, at La Jolla, reveal that the vast Canadian shield appears to be a core or nucleus for the entire continent, and that it is over 2.5 billion years old. Wrapped about the Canadian Shield are belts of ever younger rocks. At the outside, the granitic foundation rocks under the North American coasts are the youngest of all, less than 0.6 billion years of age. The continents, then, seem to be accretions of light minerals deposited by subocean conveyor belts.

Like Grimes's elliptical oceanic currents, magmatic convection cells might form a geometrical pattern that would explain the present shapes and locations of the continents. In 1944 the Dutch geophysicist F. A. Vening Meinesz suggested an arrangement of four triangular cells (remember Green's tetrahedron) with upwellings in the southern, oceanic hemisphere of the earth and the descending areas under the northern hemisphere continents. Vening Meinesz supposed that the continents had been created by accretion, but

RECAPITULATION OF SOME OCEAN-BASIN THEORIES

Theory	Catastrophic	Uniformitarian	Both	Contracting earth	Expanding earth	Neither	Differentiation of C and OB *	Shape and location of C and OB	Topography of C and OB	Frequent geologic revolutions	Accepted	Held in abeyance	Rejected
METEOR OR MOON REMNANT THEORY	X					X	X		X	X			X
METEOR CRATER THEORY	X					X	X		X	X			X
MOON BIRTH: VENUS THEORY	X					X	X	X	X				X
MOON BIRTH: TIDAL THEORY	X					X	X	X	X			X	
DRIFT AND SEA-FLOOR SPREADING		X			X			X	X		X		
JOLY CYCLIC THEORY			X	X	X				X	X			X
CHEMICAL AND GRAVITATIONAL SEPARATION	X					X	X				X		
ELLIPTICAL CURRENT THEORY	X					X	X	X					X
TETRAHEDRAL EARTH THEORY	X			X				X	X				X
BUBBLE COLLAPSE FEATURES	X					X							X
EARTHQUAKE COLLAPSE FEATURES	X					X							X
"PUDDLES" ON ROUGH EARTH		X				X							X

BASICALLY: Catastrophic / Uniformitarian / Both
DEPENDS UPON: Contracting earth / Expanding earth / Neither
ACCOUNTS FOR: Differentiation of C and OB * / Shape and location of C and OB / Topography of C and OB / Frequent geologic revolutions
GENERALLY: Accepted / Held in abeyance / Rejected

* C and OB = continents and ocean basins

27

his convection cells could also have rafted floating continents to their present locations. The triangular network created by the edges of the convection cells where the upwardly flowing magma diverges should strain the earth's crust, forming worldwide belts of volcanoes and earthquake-prone regions. This is just what is observed.

So far, emphasis has been on the formation and transportation of the continents rather than ocean basins. The inference is that all the action is continental and the sea floors are pretty dull places. Although we may think of sea floors only as those places where there are no continents, modern oceanographic ships bring back evidence that the sea floors are young, hot, and in motion. The modern view is that the ocean basins are fresh, active wounds in the planet's crust. Two of the most exciting topics in geophysics today are continental drift and sea-floor spreading. The subjects are tied together, of course, because if new sea-floor area is being created from solidifying, upwelling magma, the continents, which seem to float in the stuff, have to move elsewhere, and perhaps they are moving today. The next two chapters are consigned to these subjects. The important fact here is that the ocean basins are not ancient and static; they seem to be the youngest, solid parts of the earth's crust. The Atlantic Ocean may not even have existed 150 million years ago—yesterday, by geologic time standards. Instead of the cooling, wrinkling, dying earth of yesterday, today's geophysicists work with an expanding, dynamic earth—a planet where nature's work is not finished.

READING LIST

BULLARD, E.: "The Origin of the Oceans," *Scientific American*, Sept. 1969.

DIETZ, R. S.: "Continents and Ocean Basin Evolution by Spreading of the Sea Floor," *Nature*, June 3, 1961.

ENGEL, A. E. J.: "Geologic Evolution of North America," *Science*, Apr. 12, 1963.

FAIRBRIDGE, R. W., ed.: *The Encyclopedia of Oceanography*, Reinhold Publishing Corp., New York, 1966.

FAUL, H.: "Tektites Are Terrestrial," *Science*, June 3, 1966.

MATHER, K. F., ed.: *Sourcebook in Geology, 1900–1950*, Harvard University Press, Cambridge, 1967.

MENARD, H. W.: "The Deep Ocean Floor," *Scientific American*, Sept. 1969.

MULLER, P. M., and SJOGREN, W. L.: "Mascons: Lunar Mass Concentrations," *Science*, Aug. 16, 1968.

O'LEARY, B. T., CAMPBELL, M. J., and SAGAN, C.: "Lunar and Planetary Mass Concentrations," *Science*, Aug. 15, 1969.

SHEPARD, F. P.: *Submarine Geology*, Harper & Row, Publishers, New York, 1963.

WEGENER, A.: *The Origin of Continents and Oceans*, Dover Publications, Inc., New York, 1967 (paper).

3

RIDGES, RIFTS, AND TRENCHES

As Europe's great sea explorers fanned out over the world's seas during the sixteenth and seventeenth centuries, the surface map of the world gradually assumed its present contours. Coastlines, tropical archipelagoes, and the icy fringes of Antarctica were all duly recorded. But these explorers missed some of the most awe-inspiring features of our planet. Beneath the seas they sailed lies a submarine mountain chain that seems to girdle the world—it may be forty thousand miles long. Running along the crest of this ridge is a great crack or rift that makes the earth appear to be bursting at its seams. The early explorers also missed the submarine trenches, some of which are a mile deeper than Everest is high.

For surface ships to discern submarine topography, some sort of depth finder is required. For nearly four thousand years the standard depth finder was the sounding line. When a sounding line is used today, it is jocularly called a "Phoenician Fathometer" after the ancient seafarers who also used it. Columbus is reported to have used a four hundred fathom (2,400-foot) sounding line during his 1492 voyage. In 1818 Sir John Ross sounded to six thousand feet in Baffin Bay. But most soundings were shallow and very spotty. Hours of hard work with heavy rope were required for a single deep sounding; and even then the ship's drift and the difficulty of telling when the weight hit bottom cast doubt upon the data. No overall structural patterns could be observed for the ocean floor. Piano wire, motor-driven reels, and automatic line-measuring devices were introduced in the late 1800's; these increased the quality and frequency of deep-ocean soundings. It was the invention of the echo sounder that revolutionized sea-floor mapping. During the 1920's and 1930's echo sounding became a well-developed art in oceanography. With better topographic maps the outlines of colossal submarine geologic features began to emerge. It was a disturbing map that swam into focus.

Prior to 1900 most geologists had conceived of the ocean floors as featureless, dreary places. Their thinking seemed sound: there was little erosion in the depths, ergo there could be no Grand Canyons; neither was there enough sedimentation to build up the mile-thick strata seen on land. On the other hand there must be a few mountains because Ascension Island, the Hawaiian Islands, and other above-water hints of roughness could not be ignored. But, in general, limited soundings showed that the ocean bottoms were unexciting. The term "abyssal plain" seemed to typify the usual sea-floor situation.

The first hint that all might not be serene on the bottom came from the careful surveys of the cable-laying ships that crept across the Atlantic unreeling their cargoes in the late 1800's. The famous British *Challenger* Expedition (1872–1876) also took a detailed look at the Atlantic floor with hemp rope and heavy weights. It soon became clear that a mountain range of sorts existed between Europe and America. This was plainly a departure from the abyssal-plain view, but the idea of a general, gradual upwarping of the Atlantic floor was not particularly disturbing. The newly discovered ridge was just another antisyncline (upswelling), due to compression of the crust as the planet cooled.

The echo sounder completely demolished the concept that the ocean bottoms were merely larger versions of Bonneville Salt Flats. From 1925 to 1927 the German oceanographic ship *Meteor* criss-crossed the Atlantic making echo soundings. The mid-Atlantic "upwarping" was found to be a long mountain chain comparable in roughness and size to the Rockies. From Iceland to Tristan da Cunha, in the South Atlantic, the ridge broke through the surface on occasion, but it was now obvious that the Atlantic waters obscured one of the major features of our globe. The mid-Atlantic Ridge splits the Atlantic basin into narrow eastern and western troughs that bear no resemblance to the circular meteor craters proffered by Gilvarry in the preceding chapter.

The mid-Atlantic Ridge seemed so persistent from one end of the Atlantic to the other that the German geologist L. Kober postulated in 1928 that it was part of a world-circling mountain chain that was still in an embryonic stage. Kober's chain was the focus of earthquakes as the mountain-making process heaved up huge slabs of crust. Kober's hypothesis was the antithesis of that proposed by the Atlantis buffs, who quickly interpreted the ridge in their own way—that is, the mid-Atlantic Ridge as the foundered continent from which Atlantean kings had once ruled the world.

Since World War II all of the world's oceans have been mapped in detail by high-precision echo sounders. The global system of undersea ridges and the long encircling chain of Pacific trenches would have profoundly startled the scientists aboard the *Challenger* in 1872. These two cardinal features of the earth's crust—one protruding, the other intruding—may actually be different external expressions of a common subterranean mechanism. To understand the genesis of the ridge-trench hypothesis, a few more details about the ridges and trenches are required.

The midocean ridge system is often called the largest single geologic structure on earth. The ridge is broad—more than one thousand miles wide at most places—and generally has an elevation of between one-half and two miles above the adjacent ocean floors. Its total area as we follow it from one ocean to another is equal to that of all the continents taken together.

Echo soundings taken along courses athwart the ridges reveal a rough fractured structure, with a spine displaced several miles in places by crosswise faults that make sharply defined kinks in the ridge's backbone. The central rift or crack, the most startling feature of the ridge's structure, was discovered in 1953 by Marie Tharp, at Columbia's Lamont Geological Observatory, when she was making topographic sketches of the mid-Atlantic Ridge from echo-sounding data. With few exceptions the entire length of the ridge is cleft by this deep chasm that dwarfs the Grand Canyon. In the Atlantic the rift has a floor that averages six thousand feet below the high ridges on either side. The width of the rift ranges between eight and thirty miles. Like the ridge itself, the rift is offset where transverse faults occur. Most of the shallow (twenty-mile deep) earthquakes that occur in the Atlantic are associated with

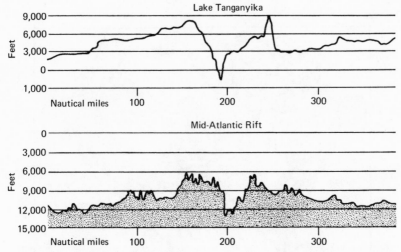

Profile of the Mid-Atlantic Ridge showing the central rift. (The vertical scale is greatly exaggerated.) The worldwide rift system cuts into several continental areas, as illustrated for Lake Tanganyika, in Africa.

the rift. Evidently, the rift in the mid-Atlantic Ridge is a basic flaw in the earth's crust, and instead of healing, the wound seems to be worsening.

By the mid-1950's seismograph stations around the world had sketched out a long locus of earthquake centers that extended down the mid-Atlantic Ridge, around the world twice, and into all oceans. In 1956 Bruce C. Heezen, a key figure in Columbia's exploration of the mid-Atlantic Ridge, predicted that echo soundings would show that the entire forty thousand mile line of earthquake centers coincided with a rifted system of ridges, just as it did in the Atlantic.[*] In other words the earth was cracked extensively, much like a ripe cantaloupe that has been dropped.

Naturally, considerable opposition developed to this dramatic picture of a wounded planet—the vision of a cracked earth is hardly Uniformitarian. The opposition first claimed that the rift was only a local phenomena and that it even disappeared along sections of the mid-Atlantic Ridge. It does indeed become hard to

---

[*] As mentioned earlier, the German Kober anticipated Heezen in 1928, though he did not know of the rift of course. There are so many hypotheses and hypothesizers in geology that true originality is hard to claim.

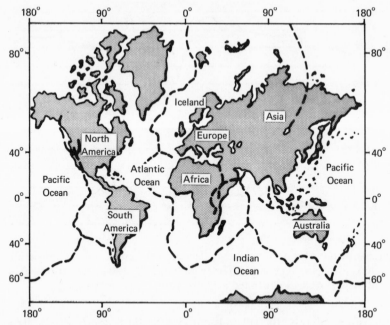

The midocean rift system (solid line) based on soundings and earthquake records. The regions where deep trenches occur are indicated by dotted lines.

trace in places, but echo soundings fully support the contention that the rifted ridge is a true worldwide structure.

The fracture in the earth's crust is not confined beneath the sea. It surfaces in Iceland, where geologists have long wondered about the Great Graben (crack) that splits the island and seems to be slowly widening. The crack also runs up along the Gulf of California to become the famous San Andreas fault that gives Californians the jitters. It then dives under the sea again to resurface as the Lynn Canal of the Alaska panhandle. The Red Sea and the perplexing Rift Valleys of east Africa also seem to be extensions of the worldwide rift. Wherever the rift goes, it brings earthquakes. Wherever geologists see the rift invade the land, they see evidence of great tension in the earth's crust.

There is, however, one region of the globe that the rift seems to avoid: the north central and northwestern Pacific Ocean. As we know, the Pacific Ocean basin is quite different from the other ocean basins and may have had a different origin. Many of the Pacific earthquakes originate in regions near the deep-sea trenches that fringe eastern Asia and South America.

From the geophysicists come some other facts that may help us guess the origin of the midoceanic ridge and its central rift. A large positive magnetic anomaly is associated with the ridge; that is, the earth's magnetic field, measured by ships passing over the ridge, rises sharply above the values measured on either side. The magnetic anomaly could indicate subterranean differences in rock composition. There is also a moderate negative gravity anomaly along the rift. By detonating small charges of explosives and noting how the sound waves are refracted in the ridge-rift structure, geophysicists have concluded that the composition of the crust in the vicinity of the rift is intermediate between the main ocean-floor crust and the earth's mantle that underlies all of the ocean floors and continents. Heat-flow measurements along the floor of the rift have demonstrated that several times more heat flows through the crack than through the ocean floor proper. It is hard to resist leaping to the conclusion that hot plastic mantle material is extruded along the rift, forming the oceanic ridges.

Direct physical contact with the rocks of the ridge complex has not been mentioned. In recent years many rocks and sediment samples have been dredged up from the midoceanic ridges, especially in the Atlantic. Cores have also been extracted from the surface rock structure. The oldest indigenous sedimentary rocks brought up from the mid-Atlantic Ridge are from the Miocene period, only twelve to twenty million years ago. (A few extraneous rocks found beneath the sea may have been carried there as ballast by ships, by ice floes, and even by kelp beds.) The volcanic rocks disinterred from the depths are also young, fewer than five million years old. The crests of the ridges paralleling the rift seem to be built of fresh basalt and are not covered by sediment. Farther down the flanks of the ridge away from the rift, more and more sediment is found as the ocean floor is approached. The entire ridge is apparently a rather young structure, but its flanks are older than its crest. Within the rift itself the basalt walls appear to have been metamorphosed by heat and pressure. The floors of the rift valleys have light coatings of sediment that could have accumulated in just a hundred thousand years or so—almost yesterday by geological reckoning. A final fact heightens the ridge-rift mystery;

coarse sands typical of continental sediments have been dredged up from the mid-Atlantic Ridge.

The echo sounder was also instrumental in sketching out the features of a system of great crevasses that ring the Pacific basin. Not to be confused with the submarine canyons that are incised in the continental shelves and slopes, the trenches are true ocean-floor features, reaching almost seven miles below sea level down into the basic fabric of the earth's crust. In contrast to the submarine canyons, which seem to be scoured out of offshore sediment and continental rock by erosion, the trenches appear to have been created by large-scale deformations of the earth's crust. The trenches are also quite different from the planet-encircling rift system. The floors of the rift system are above the level of the surrounding basin, but the trenches are sharp, V-shaped wrinkles that penetrate well below the level of the ocean basin.

Most of the deep trenches fringe the Pacific Ocean, except for the North American West Coast, where the ocean rift system comes ashore to become the famous San Andreas fault. The almost legendary trenches of the western Pacific are the deepest; viz., the Mariana Trench, 36,000 feet deep and 1,600 miles long, and the Tonga Trench, 35,400 feet deep and 870 miles long. The Atlantic boasts only a few trenches, such as the Puerto Rico Trench and the Cayman Trench, both much shallower than their Pacific counterparts. Most trenches are associated with arcs of islands, such as the Marianas and the West Indies, or with young coastal mountain ranges. They are not features of the central ocean basins; they are fringe structures, albeit they do not form continuous or orderly arcs.

Like the rifts, the trenches are the epicenters of many earthquakes. The trenches also parallel strings of volcanoes in the Pacific. One also finds strong negative gravity anomalies associated with them, once again like the ocean rift. No unusual magnetic anomalies have been discovered, however; neither does the rate of heat flow outward from the crust seem unusual. In these last two aspects, the trenches differ markedly from the rifts.

In January 1960 the bathyscaph *Trieste* carried Jacques Piccard and Lieutenant Don Walsh to the bottom of the Mariana deep, almost seven miles beneath the surface. Like most other trenches the Mariana Trench was found to have a narrow, flat bottom a couple miles across. The floor was evidently created by the slow accumulation of diatomaceous ooze. The trenches seem considerably more ancient than the rifts. Each trench seems to have its own peculiar fauna, indicating that these animals have long been trapped and isolated in their own trenches.

Why discuss trenches and rifts in the same breath if they are so different? They may be different geologic structures but still be related through a common causative mechanism. Taken together, all evidence hints at an upwelling of rocky material along the rift system and a corresponding engulfment of similar material in the vicinity of the trenches. We may legitimately use these hints to construct a model based upon upwelling and disappearance of terrestrial material, but no matter how intuitively satisfying the model is, it may not be supported by the facts. Nor is it essential that both trench and rift be explained by the same hypothesis.

The model alluded to is called the "sea-floor spreading" model; and it was originally based upon the idea of the subterranean convection cell introduced in the last chapter.

The sea-floor spreading hypothesis is basically a Uniformitarian idea because it involves a slow process observable in the field today. However, it represents such a departure from standard geological thinking that the conventional geologist could hardly be expected to originate such a heretical idea. For the first half of this century, then, the accepted view held that the mid-Atlantic Ridge was an expression of the same forces that mold the continental mountains; that is, compressional forces arising from the earth's shrinking and volcanic action. As always a few scientists swam against the current.

Two nonconformists, John Joly and Alfred Wegener, were early proponents of sea-floor spreading. In his 1925 book, *The Surface History of the Earth,* Joly perceived that his radioactively heated expanding earth must develop rifts through which hot magma would ascend and freeze. The solidified crust would gradually be forced away from the rift by new upwellings of magma. In the process new ocean floor would be created. And, very significantly, the new floor would be radically different from continental crust—just as we observe. Joly also recognized the Red Sea cleft and the African Great Rift Valleys as further evidence of great tension in the earth's crust. Alfred Wegener was one of the greatest proponents of drifting continents. Of course, continents cannot drift apart without new crust forming in their wakes. Wegener took readily to Joly's

hot, expanding earth approach, and in the 1924 edition of his classic, *The Origin of Continents and Oceans,* he stated:

> If the basalt layer under the granite were really specially fluid as as-
> sumed, then, as the Atlantic rift opened wider progressively, this layer
> would have had to rise up here, subsequently flowing steadily out
> from both sides; it would first have formed the whole ocean floor, and
> would still today form the greater part of it.

Wegener was able to draw on the work of several other scientists.
The Austrian geologist Otto Ampferer had suggested as early as
1906 that the viscous drag of thermal convection currents under the
crust was an important source of horizontal forces on the continents. In the late 1920's the German G. Kirsch, and the English-
man Arthur Holmes studied subterranean convection cells as movers of continents. Holmes is regarded as the father of the "sea-floor
spreading" concept.

The "modern" hypothesis of sea-floor spreading was developed
by the American geophysicist Harry H. Hess and published in
1962. Fundamentally it is the same as that suggested by Wegener
and Holmes: upwelling magma creates tension in the sea floors;
light continental blocks are rafted to down-welling sites where
their positions are stabilized, and new sea floor is formed as
magma solidifies in the regions of the rifts.

Sea-floor spreading is an attractive mechanism that allows scientists to place continents where their theories about geologic and
biologic evolution require. But is the hypothesis susceptible of
proof in the field? The hypothesis calls for spreading rates of just a
few centimeters per year. How can such slight motion be measured, particularly far down in the blackness of the midocean rifts?
The distance between Europe and North America could be measured to see if they are receding from each other, but even with
modern geodetic satellites to help, our instruments lack the necessary precision. The best spots to detect sea-floor spreading would
be where the midocean rifts apparently rise to the surface, say,
along Iceland's Great Graben or the San Andreas fault in California. There is plenty of geologic evidence at such spots supporting
growth of the rift (Iceland) and slippage parallel to the fault (California), but no one can measure the motion directly. Today lasers
give us a tool with the ability to measure minute movements across
such faults, but no significant results have yet been obtained. In

any case displacements along faults prove only that the earth's crust is not in equilibrium and not necessarily that the sea floor is spreading.

For the present, at least, the most convincing evidence supporting sea-floor spreading comes from geologic clocks—clocks that measure time in millennia rather than in man's short life. One rather convincing time gauge is the sediment thickness on the flanks of the midocean ridges, especially the mid-Atlantic Ridge. As already mentioned, little sediment occurs around the rift proper at the ridge crest, but sediment thickens with distance from the rift. Under a steady rain of deep-sea sediment, a sediment pattern such as this implies the continuous creation of new sea-floor material in the neighborhood of the rifts.

Recent sea-bottom cores taken from the East Pacific Rise, where the rift strikes northward toward the Gulf of California, show the same pattern of sediment thickness. Further, the thin sediments near the crest are from the Pleistocene period (beginning about 600,000 years ago) while the thicker sediments out on the flanks contain pre-Pleistocene sediments. Sea-floor spreading, if it really occurs, continues right down to recent geologic times.

A more fascinating kind of geologic clock also supports sea-floor spreading. With a "magnetic" clock even the rate of sea-floor spreading can be measured. The clock's action depends upon the fact that volcanic rocks—including the sheets of magma supposedly oozing out along the midocean ridges—become magnetized along the direction in which the earth's field pointed at the time they cooled and solidified. This fact might rightly be consigned to the "so-what" file if the earth's magnetic field has remained fixed throughout geologic time. However, back in 1906, Bernard Brunhes, a French physicist, discovered that some volcanic rocks were magnetized in a direction opposite to that of the present field. Undeterred by the doctrine of Uniformitarianism, Brunhes concluded that the earth's field must have reversed. The discovery was confirmed wherever volcanic rocks were found. The news fell mostly on deaf ears. Paleomagnetism has long been employed to prove that the earth's poles have wandered and that continents might have drifted. However, the idea of complete reversal of magnetic polarity seems to have been hard to swallow.

Facts now indicate that the earth's magnetic poles have actually reversed positions several times within the last several million years. This can be shown by taking cored samples of volcanic rock, measuring the directions of magnetization, and then establishing ages by measuring the amount of potassium-40 that has decayed to gaseous argon-41. By comparing many hundreds of samples taken the world over, geophysicists have constructed a magnetic calendar, showing reversals at about 0.7, 2.4, and 3.35 million years ago. The time periods between these reversals are termed "magnetic epochs." The epochs alternate normal and reversed period of magnetic polarity. Several pairs of short-term reversals called "events" are superimposed upon the epochs (see illustration).

Cores extracted from the volcanic rocks on each side of the mid-ocean rifts reveal patterns of remarkable symmetry and linearity. Parallel to the central rifts, linear zones of normal and reversed polarity rocks stretch for hundreds of miles along the sea floor. The

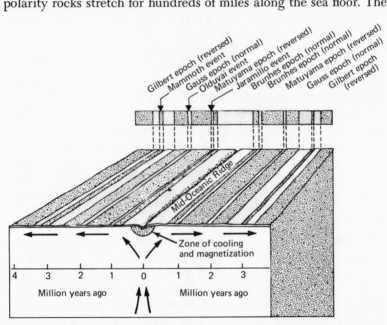

Sheets of cooling volcanic rock issuing from the midocean rifts apparently record the polarity of the magnetic field existing at the time they solidify. The linear patterns, which are symmetrical about the rift, can be dated by radioactive methods, giving us a tape recording of the earth's magnetic history. Superimposed on the main epochs are short events: the Jaramillo and Olduvai events (Matuyama epoch) and the Mammoth event (Gauss epoch).

farther the strips are from the rift, the older their radioactive ages. Linearity and symmetry have been found in the vicinities of all rifts. By measuring the width of the strips, sea-floor spreading rates of a centimeter or two a year are derived. The consistency and universality of these measurements give the sea-floor spreading and continental-drift hypotheses credence they did not enjoy prior to 1960.

The magnetic data are neat and convincing. Scientists, though, are always suspicious, as they should be, with new hypotheses and data that have not yet withstood the test of time. The first impulse is to question the stability of the polarization of magnetized volcanic rocks. Possibly external conditions—heat, for example, or even lightning (on land)—might have reversed the polarities of the terrestrial samples used to establish the time scale and the anomaly pattern about the midocean rifts. Laboratory experiments, however, have shown that the magnetic stability of volcanic rocks seems too strong to be disturbed significantly by external forces. Another criticism involves the concept of magnetic self-reversal, a term describing the tendency of some rocks to become magnetized with reversed polarity while others nearby assume normal polarity in the same external field. In other words some kinds of rocks are nonconformists. Again, laboratory tests came to the rescue. Tests with hundreds of samples of volcanic rock proved that less than 1 percent were self-reversing. Other scientists object that the heat-flow patterns around the midocean rifts are not exactly what one would expect from the sea-floor spreading model nor are they the same from one rift to another. There are similar problems with details near the ocean trenches where the hot fluid magma, flowing under the ocean floor it created in the geologic past, finally turns down and disappears into the depths. At present, however, the evidence generally supports the sea-floor spreading hypothesis, which, in its most popular form, depends upon a hot expanding earth with internal convection cells.

Egon Orowan, a professor of mechanical engineering at Massachusetts Institute of Technology, has recently proposed a new source for the horizontal forces that promote sea-floor spreading. Orowan views the mid-Atlantic Ridge as a welt or swelling that rises above the ocean because the subterranean olivine rock along the rift in the crust has absorbed seawater to form serpentine. Olivine can take up to one-fourth its volume in water during the serpentinization process. As the ridge of water-expanded rock rises around the mid-Atlantic Rift, horizontal forces are generated perpendicular to the centerline of the rift. The forces arise as the ridge tends to flatten out by "flowing" glacierlike toward the Americas on one side and Europe and Africa on the other. Orowan calculates that the forces thus created are sufficient to push the continents apart several inches a year. As the substance of the ridge flows east and west (as a plastic solid not a liquid), new olivine rises from the crack and is serpentinized by seawater to perpetuate sea-floor spreading. In effect the convection cells powered by the heat of radioactive decay have been replaced by a process that first turns chemical energy (from serpentinization) into gravitational energy, and, finally, into propulsive energy for sea-floor spreading and continental drift.

The supporters of the continental-drift hypotheses—mainly those scientists attempting to explain paleomagnetic patterns or the global distribution of animal species—have been delighted with the sea-floor spreading hypothesis. It has provided them with a reasonable mechanism, a sort of conveyor belt that transports continents from one point to another. Indeed, the theory of continental drift might be dead today without the stimuli of paleomagnetic data and the sea-floor spreading hypothesis. In the following chapter the whole concept of continental drift will be scrutinized, but already the earth seems a plastic living thing.

The overtones of Catastrophism sound stronger and stronger as the symphony continues. Some jarring notes were recorded by the magnetic anomalies on the ocean floors. Where did the discordances originate? We know so little about our enshrouding magnetic field. The wonder is that it exists at all! The poles wander, and there is evidence that they reverse positions on occasion. The only real theory of the earth's magnetic field attributes it to the slow ponderous motion of convection cells under the mantle. How could these be reversed to change the fields' polarity? The mechanism is not known for certain; but when more venturesome scientists stand back to view the entire panorama of the earth's history, they note disturbing coincidences. For example, the last reversal of the earth's field seems to coincide with the age of the tektites, those fragments of molten glass strewn from the Indian Ocean to the South Pacific. The implication is that some cosmic body careened into the earth blasting out the molten matter that formed the tektites and somehow reversing the earth's magnetic field at the same time.

READING LIST

COX, A. *et al.:* "Reversals of the Earth's Magnetic Field," *Scientific American,* Feb. 1967.

FAIRBRIDGE, R. W., ed.: *The Encyclopedia of Oceanography,* Reinhold Publishing Corp., New York, 1966.

HEEZEN, B. C.: "The Rift in the Ocean Floor," *Scientific American,* Oct. 1960.

——"Tektites and Geomagnetic Reversals," *Scientific American,* July 1967.

HESS, H. H.: "The Evolution of Ocean Basins," in *Petrological Studies,* A. E. J. Engle *et al.,* eds., Geological Society of America, 1962 (The classic reference on sea-floor spreading).

JOLY, J.: *The Surface History of the Earth,* Clarendon Press, New York, 1925.

MENARD, H. W.: "The Deep-Ocean Floor," *Scientific American,* Sept. 1969.

OROWAN, E.: "Age of the Ocean Floor," *Science,* Oct. 21, 1966.

——"The Origin of the Ocean Ridges," *Scientific American,* Nov. 1969.

VINE, F. J., and MATTHEWS, D. H.: "Magnetic Anomalies over Oceanic Ridges," *Nature,* Sept. 7, 1963 (The classic reference on magnetic anomalies).

——"Spreading of the Ocean Floor: New Evidence," *Science,* Dec. 16, 1966.

WEGENER, A.: *The Origin of Continents and Oceans,* Dover Publications, Inc., New York, 1966 (paperback).

4

CONTINENTAL DRIFT

With our narrow perspectives in time and space, it is hard to imagine the high-floating continents sailing stately, aeon by aeon, across hidden subterranean seas of viscous basalt. In their wakes they leave festoons of islands and a thin pavement of youthful ocean floor still straining and cracking under surging convection cells of magma or other forces that set the whole machine in motion. The vision has grandeur—no question about that—but evidence for and against the concept is largely indirect. Pierre Termier, a French geologist, once called continental drift "a beautiful dream, a dream of a great poet. One tries to embrace it, and finds that he has in his arms but a little vapor or smoke . . . ."

Termier said this almost a half century ago. Today his vapor has more substance. In fact, in the April 1968 issue of *Scientific American*, Patrick M. Hurley entitled his analysis of the situation "The Confirmation of Continental Drift." Many scientists, especially the geologists, have not yet embraced continental drift as firmly and energetically as Hurley's title implies. Yet, a mere decade ago the world of science had written the epitaph for continental drift. Continental drift was placed contemptuously on a par with the concepts of phlogiston and the all-pervading ether—all the other blind alleys of science.

Continental drift grates against the intuition of the conventional geologist, whose thinking has been shaped by generations of professors who taught that this planet's crust had been frozen solid for aeons. Certainly our "sense of terra firma" seems to deny that we live on careening continents; neither was it possible a decade ago to conceive of a natural force great enough to propel such gigantic crustal blocks across a tacky mantle covered with a frozen crust several miles thick. But indirect evidence, particularly from paleomagnetism and paleobotany, strongly implies that some of the earth's land masses were once united and have since drifted apart. There is also supporting evidence from the disciplines of geology, geophysics, and climatology, which eager proponents of continental drift (the "drifters") weave deftly into a convincing story.

Even with our sophisticated magnetic analyses of rocks and the century-long search for transoceanic correlations of fauna and flora, the most appealing observation supporting continental drift is the parallelism of coastlines on opposite sides of the Atlantic. South America and Africa simply look as though they might fit together, North America and Europe, too, but with a poorer fit. And they do, especially when a few thousand feet of seawater are siphoned off and the edges of the continental shelves are compared.

Even Francis Bacon, back in 1620, suggested that the Americas, Europe, and Africa had once been united. Von Humboldt offered the idea once more in 1800. Again and again, geologists and non-geologists, professional scientists and amateurs, have played with the jigsaw puzzle of the continents—the coastal similarities across the Atlantic are too strong to resist.

In the 1850's Antonio Snider, an American gentleman residing in France, noted the similarities between American and European fossil plants of the Carboniferous period (about 300 million years ago). In his 1858 book, *The Creation and Its Mysteries Revealed,* he suggested that all the continents had once been joined together and that a great crevasse had later opened between the Americas on one side and Europe and Africa on the other. Thus Snider was one of the first to transcend the jigsaw-puzzle argument with paleontological data. Perhaps Snider went too far because he also discussed similarities in human culture around the Atlantic shores and their possible relation to the legend of lost Atlantis.

Eduard Suess, the great Austrian geologist, took continental drift a step further at the turn of the century. Suess pointed out that one distinctive geologic formation from the Permian period, containing certain plants, shells, amphibia, and reptiles, was found in India, Australia, and other lands in the Southern Hemisphere. Particularly impressive were the extensive coal measures found in these southern lands, which were dominated by the distinctive tongue-shaped leaves of the Permian *Glossopteris* flora. These plants were extremely rare in the Northern Hemisphere. The southern continents, he reasoned, must have been joined when these closely correlated sediments were deposited. He called the ancient continent Gondwanaland, after the name of the geologic province in east central India. Late in 1967 geologists working in Antarctica discovered the fossilized bones of the first land vertebrate ever found on that continent. This amphibian, a laybrinthdont, was prevalent throughout Suess's Gondwanaland and would have greatly strengthened his case had he known about it.

But why should the continents drift in the first place? Transoceanic correlations say nothing about physical mechanism. Some of these early drifters invoked the centrifugal and Coriolis forces that affect the atmosphere and oceans, but these seemed too weak to rend whole continents. The most interesting mechanism was suggested by the Reverend Osmond Fisher (in 1882) and William H. Pickering (in 1907): the Americas were drifting westward to fill up the void created when the moon was flung off the earth and created the Pacific basin. The mid-Atlantic Rift, which seems to be busily manufacturing new ocean floor, and the broad Pacific basin are both manifestations of the same catastrophe according to this view. (The proposed rates of drift and sea-floor spreading [a few inches a year] would be adequate to move North America three thousand miles from Europe in 300 million years.)

All the surmise and speculation made little impression on orthodox geology until just before World War I, when two geologists, an American, Frank B. Taylor, and a German, Alfred L. Wegener, amassed so much evidence favoring continental drift that scientists had to take sides. Wegener's work was the most comprehensive and convincing. In taking a hard line on drift, he founded the Wegenerian school of thought. Wegener stated the hypothesis so clearly and precisely that facts, as well as people, had to be either for or against continental drift. In science a hypothesis is useful if it stimulates arguments and if it is susceptible to proof or disproof. Wegener's formulation of the continental drift hypothesis certainly stimulated science; but, as we shall see, his edifice soon became the target of the doubters who began accumulating contradictory evidence at a rapid clip. To the orthodox, drift became a bad word, and Wegener was at the center of the controversy.

Born in Berlin on November 1, 1880, the son of an evangelical preacher, Alfred Lothar Wegener began his career as a meteorologist. His interest in the polar air masses, which are so critical to weather, led him to join the 1906 Danish expedition to northeast Greenland. He soon became a specialist on Greenland and made several expeditions to this frozen land mass.

In 1910, while teaching at the Physical Institute, at Marlburg, Germany, he was impressed by the congruency of the Atlantic coastlines and the possibility of drift, as had so many others before him. At the time he considered actual drift of the continents unlikely. Wegener describes in *The Origin of Continents and Oceans* what happened then as follows:

> In the autumn of 1911 I became acquainted (through a collection of references which came into my hands by accident) with the paleontological evidence of the former land connection between Brazil and Africa, of which I had not previously known. This induced me to undertake a hasty analysis of the results of research in this direction in the spheres of geology and paleontology, whereby such important confirmations were yielded that I was convinced of the fundamental correctness of my idea.

Another fact that impressed Wegener was the *apparent* slow westward motion of Greenland—about fifteen feet per year—according to the crude longitude measurements of those days.

After thorough study Wegener proposed in 1912 that all the continents had originally formed a single mass, which he called Pangaea or "All-earth." Pangaea was surrounded by Panthalassa or "All-sea." Pangaea was much larger than Suess's Gondwanaland because it also included the northern continents. Under the action of gravitational force acting upon slanting, protruding blocks of continental crust, Pangaea eventually sundered and the continents floated freely upon the magma sea. These horizontal forces might be compared to those acting upon a wooden block sliding down an inclined surface. Later, just prior to his death, Wegener recognized that internal convection forces might push the continents horizontally. Whatever the source of the energy, Wegener visualized the freed continents pushing into the viscous magma, throwing up mountain ranges along their bows (the Andes and Rockies) as the rocky sea offered resistance. In their wakes were strung island arcs, such as the West Indies and Japan. Of course, great climatic changes—which geologists had recognized but hitherto could not explain convincingly with conventional theories—ensued as the continents changed position.

Wegener's first papers on the subject of continental drift were published in 1912. In 1915, making use of a prolonged sick leave from war service, he completed his classic work *Die Entstehung der Kontinents und Ozeane,* which was published by Vieweg in 1915. The first English edition of *The Origin of Continents and Oceans* appeared in 1924. With the recent revival of interest in continental drift, a paperback edition has appeared. It is again fashionable to be a Wegenerian—or at least a modified Wegenerian.

Wegener marshaled his arguments supporting continental drift in orderly Germanic array in the several editions of his book: geodetic arguments, geophysical arguments, geologic arguments, paleontologic arguments, biological arguments, and paleoclimatic arguments. Many of his analyses are still valid and will appear in the overall assessment of continental drift later in this chapter.

The Wegener hypothesis of continental drift, meticulously formulated and supported as it was, was embraced immediately by biologists and paleontologists urgently in need of either mobile continents or land bridges to explain the distribution of species through geologic time. Geologists and geophysicists, however, received the theory with disdain. Wegener's forces were too puny to shove whole continents through the ocean crust; his thoughts about physical changes of the earth's axis of rotation were ridiculous; when he proposed subcrustal convection currents as the prime movers of continents, he piled nonsense upon nonsense—at least in the eyes of most geologists.

Even before Wegener disappeared forever during his fourth Greenland expedition in 1930, the hypothesis of continental drift began to wane in importance and the heavy hand of doctrinaire Uniformitarianism descended upon scientific thinking about the earth's past. Although continental drift is slow and noncatastrophic, most true Uniformitarians considered mobile continents an affront to the spirit of Uniformitarianism. They felt that the evidence marshaled by Wegener could be explained by temporary land bridges between continents without invoking a hypothesis as un-conventional as continental drift. As usual, a few brave souls swam against the current of prevailing opinion. Two deserve special mention.

The first is Alex L. du Toit, who was a professor at Johannes-burg University, in South Africa, during the dark ages of the conti-nental drift theory. Like many other scientists who worked around the edges of the southern continents, Du Toit was struck by the many transoceanic geologic and biologic correlations. His 1921 paper "Land Connections Between the Other Continents and South Africa in the Past" revealed him to be an early convert to conti-nental drift. His key work, *Our Wandering Continents,* first ap-peared in 1937. It is dedicated to Alfred Wegener and bears on the title page the phrase "Africa forms the key," referring to Africa's dominant position in his supposed original configuration of the southern continents (Africa, South America, Australia, Antarctica, and parts of India). Indeed, Du Toit differed from Wegener in hy-pothesizing two primordial continents; the southerly Gondwana-land (following Suess's terminology) and the northerly Laurasia, formed from North America, Asia, Europe, and Greenland. Du Toit also emphasized the possible role of subcrustal convection currents in the propulsion of continents. Like Wegener, Joly, and a few other contemporaries, Du Toit was well ahead of his time. Take, for example, his assessment of the impact of the radical drift theory on one's outlook:

. . . the inquirer will find it necessary to reorient his ideas. For the first time he will get glimpses—albeit imperfect as yet—of a pulsating restless earth, all parts of which are in greater or less degrees of move-ment in respect to the axis of rotation, having been so, moreover, throughout geological time.

Although Du Toit believed that the earth was much more dynamic than most of his scientific colleagues believed it was, he could hardly be called a Catastrophist because he attributed the earth's restlessness to permanent, slowly acting internal forces.

Definitely on the Catastrophist side of the fence was Howard B. Baker, a little-known amateur theorist in geology. His remarkably perceptive 1932 book *The Atlantic Rift and Its Meaning* was men-tioned in chapter 1 in connection with the origin of the ocean ba-sins. Baker's earliest thoughts about continental drift appeared in

1911 and 1912 in the *Detroit Free Press* and the *Michigan Academy of Sciences Annual Report*. Of most interest here is his view that the present continents are fragments of a single primitive land mass that rushed Pacificward during the late Miocene or early Pliocene periods. In Baker's scheme of things the continents were set in motion when the planet Venus came close enough to the earth to gravitationally strip off a large section of the earth's crust, which then went to form the moon. Thus, Baker was a predecessor of Immanuel Velikovsky, although Baker relied primarily on physical evidence of Catastrophism, while Velikovsky leaned heavily on events and myths recorded in the ancient literature. Du Toit found Baker's scenario "bizarre" and "unacceptable," though of "refreshing originality." Unfortunately Baker's major works were mimeographed in very limited editions.

Baker and Du Toit, one an amateur, the other a respected professional, could not stem the decline in popularity of continental drift. Even though a most impressive mass of correlations between continents had been assembled and even though Wegener's original hypothesis had been modified and greatly strengthened, scientist after scientist went on record against the concept. It was almost as if a wave of disapproval had swept the scientific community and one had to join in the chorus. Condemnation of continental drift was *the* thing to do. The period from 1920 to 1960 marked the nadir of continental drift.

The rallying of scientists to the opposition camp was not in itself bad, for the hypothesis of continental drift did raise many problems and was supported by only indirect evidence. In other words correlations between fauna, flora, and rock formations were not enough for the skeptics. They wanted to measure directly the speeds of the continents with some sort of stopwatch. Some hard rethinking was in order. The battle, however, became violent and frequently abusive. Wegener was attacked without mercy as sloppy and incompetent. "Can we call geology a science when there exists such a difference of opinion on fundamental matters as to make it possible for a theory such as this to run wild?" stated one scientist in the late 1920's. Geologists objected that the acceptance of continental drift would require rewriting all geology books, as if this were a criterion for drift's falseness. The attacks on the theory and on Wegener himself remind one of the Velikovsky incident, mentioned in chapter 2. In both instances there is abundant indirect evidence tending to favor a bold, unorthodox hypothesis. In addition there are defects in the original formulation of the theory and considerable negative evidence. Inevitably, the result is intense, often unscientific controversy. Relativity, cosmic rays, the Copernican theory, to name only a few, have all undergone the same history of trial, ridicule, and insult and have survived and prospered. In the melodrama of continental drift, 1960 was the time to ask: who will save continental drift?

Summarizing, here are the main objections to continental drift as leveled by the doubters:

1. The convection and gravitation forces enlisted by Wegener and his supporters are far too small to push continents around. This was termed the "Atlas problem."
2. If the rocks of the sea floor slowly yielded to the prows of the moving continents, the continental rocks (the ships) must be harder and more rigid than those of the sea floor. Yet Wegener claimed that the same horizontal forces had created the mountain ranges on the continental prows. The continents could not be rigid and pliable at the same time.
3. Wegener could not explain why the continents drifted only in the last several hundred million years, a time equal to less than 5 percent of the earth's history.
4. The continental jigsaw puzzles do not really match well. All similarities are accidental and imply no general law.
5. Wegener grossly overestimated the paleontological similarities between continents.
6. Drift implies climatic changes much greater than those that actually occurred.
7. Wegener's presentation was most unscientific. He took from the literature only what suited his purposes.
8. Stable continents and ephemeral land bridges are sufficient to explain known facts.

As late as 1959 the great English scientist Harold Jeffreys, in the fourth edition of his famous treatise *The Earth,* dismissed continental drift as: "Quantitatively insufficient and qualitatively inapplicable . . . . an explanation which explains nothing which we wish to explain."

Continental drift languished until the 1960's, still cherished by the paleobotanists as a helpful working hypothesis but frowned upon by the geologists and geophysicists, particularly the latter, who disdained the "soft," nonmathematical paleontological record. It was to little avail that the paleobotanists complained that the vaunted "precise" discipline of geophysics had been pretty flexible itself when it revolutionized its whole concept of the earth several times; viz., contracting earth to expanding earth. In the eyes of the paleobotanists the geophysicists had little grounds for self-righteousness. The paleobotanists were right, because within a decade almost all geophysicists had jumped back on the drifter bandwagon. It was, in fact, the magnetic measurements of ancient continental rocks by the geophysicists that helped stimulate the renaissance of the hypothesis of continental drift.

Although Baron von Humboldt noted in 1797 that some rocks were permanently magnetized in directions different from the then prevailing magnetic field, the discipline of paleomagnetism has flourished only since the 1950's. Given a rock formation displaying permanent, stable magnetism, the scientist assumes that this field was established at the time the rock cooled and solidified and the direction of magnetization was the same as that of the earth's field. Thus, if he can determine the age of the rocks and their orientation at the time of formation, his measurement of their direction of magnetization will tell him the direction of the local magnetic field at that point in geologic history. By correlating many measurements from many sites, he can infer the positions of the earth's magnetic poles as a function of geologic time. The investigator must beware of several pitfalls, however:

1. Local magnetic anomalies (iron-ore bodies) may have severely distorted the field. The likelihood of this kind of error is reduced by making measurements over a wide area.
2. The ancient earth's field may not have been the simple dipole field we find today. Even if it were a dipole field, it might not have been aligned with the physical axis of rotation. (The two axes are more than 10 degrees apart today.) The so-called "axial dipole" assumption is critical to paleomagnetism. Although most geophysicists accept its truth, we really do not understand how our planet's magnetic field is created. If magmatic convection currents below the crust behave as dynamos, convection patterns in the past may have generated more than one pair of poles. Then, too, there may have been polar reversals in the distant geologic past like those we believe occurred within the last several million years, as inferred by the measurements of the young sea floor near the ocean ridges (chapter 2).
3. The orientation of the rocks being analyzed may have been altered by

mountain-making processes, other geologic upheavals, or, of course, continental drift. Knowledge of geologic history of the region can help correct such errors.

4. Geological ages are often imprecise. Even ages based upon radioactivity measurements depend upon several assumptions.

This list of cautions may seem to weaken any case for continental drift founded on paleomagnetic evidence. However these problems are acknowledged and many experimental precautions are taken to reduce error. When care is exercised and when many rock samples from a single continent are analyzed, they are rather consistent for each point in time. The very consistency of data based upon samples from different localities and geologic times is one of the great strengths of paleomagnetism.

When the implied positions of the north pole are plotted versus time for each continent, for the last several hundred million years, a polar "track" is traced out. There are two obvious interpretations of these tracks: (1) the poles actually did wander, perhaps as the internal convection patterns changed; or (2) the poles remained

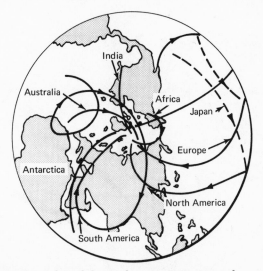

Paleomagnetic data plotted for each continent separately result in a tangle of different time tracks for the earth's north magnetic pole as it is traced backward in time. If the continents are allowed to drift back to one or two primordial unified supercontinents, the polar time tracks combine into a single line.

fixed and the continents with their burdens of previously magnetized rocks wandered. These two schools of thought were headed by two of the major modern contributors to paleomagnetism. Patrick M. S. Blackett and his colleagues at London University favored the continental-drift interpretation, while S. K. Runcorn, then at Cambridge University, liked the polar wandering explanation.

The impasse was resolved as more data became available from all of the continents. When the polar-track data for each continent were plotted on the same globe, they did not form a single track but rather separate tracks for each continent that diverged as one looked further back in time. Such divergence could be explained only in terms of wayward continents. By sliding the continents around the globe, the diverging tracks could be made to merge. The magnetic data gave drifters a road map of continental travels. By assuming fixed magnetic poles—as if one were watching a movie running backward—the continents crept from their present locations back into one or two consolidated primordial land masses. After three decades of abuse Wegener's name was cleared by a

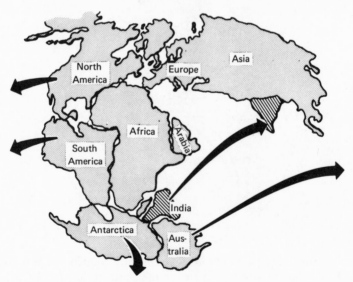

A possible arrangement of the continents before drift began some 250 million years ago. Supposedly the continents are still drifting into the Pacific void.

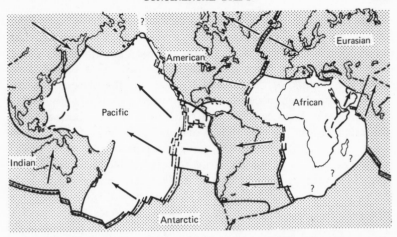

There are six major crustal plates and several smaller ones. The arrows indicate the directions of drift. The ragged edges along the ocean rift system are due to faults perpendicular to the rift (called offsets). The rift itself is dotted, while the solid parallel lines indicate relative growth rates. Heavy solid lines represent trenches where one crustal plate dives under another. (Reproduced with permission from Sir Edward Bullard, *Scientific American*, Sept. 1969, pp. 74–75)

kind of data he scarcely considered in all his laborious correlation of facts.

One of the major objections to continental drift lay in Wegener's assertion that the continents sailed like icebreakers through the sea-floor crust. The very recent sea-floor spreading hypothesis has dispelled the idea of shiplike continents. In the early 1960's it was thought that the continental blocks rode on conveyor belts of moving rock, the viscous magma flowed beneath the rocky crust, urging the sea floor and its cargo along. In the sea-floor spreading hypothesis, new sea floor issues from the midocean rifts and then disappears back into the depths along the deep-sea trenches. The conveyor-belt idea, though tentative like all hypotheses, has made the mechanics of continental drift more palatable.

The most recent "model" of the earth's crust divides the surface up into six major crustal plates and several smaller ones. The sea-bottom portion of a major plate is probably between forty and seventy miles thick; the continental portion is thicker. The plates float on the viscous mantle beneath. Crustal plates are separated by the world-circling rift system, growing outward from these "spreading centers" at rates of from one-half to five inches per year. In what has come to be called "plate tectonics," the growing plates carry the continents with them and butt up against other plates. Two things can happen at these contact points: (1) the plate edges can meet squarely, crumpling up mountain ranges on the continent (as along the west coast of South America); or (2) the sea-bottom plate slides underneath the continental plate, creating deep trenches and strings of volcanic islands (as along margins of the western Pacific). The crustal plates have not been mapped in detail as yet; even the total number of plates is in doubt. If you have ever seen ice break up on a pond or lake in the spring, you can transfer this vision of the grating, colliding, overriding ice floes to the earth's crustal plates. They, too, drift and collide, but on a time scale of millions of years rather than minutes. Conceivably, our planet's countenance has changed slowly in this way from the time the first solid crust began to form three billion years ago, but the ocean floors we see today are only 100 million to 300 million years old.

Both the drift and sea-floor spreading hypotheses rest heavily upon paleomagnetism; a rather new discipline with foundations that have yet to stand the test of time. Such basic things as polarity reversal and field generation are not understood. Actually, in view of the tremendous number of disparate facts from different realms of science that can be explained by continental drift, it might be more proper to say that continental drift supports the assumptions of paleomagnetism. Nevertheless, paleomagnetic data did remove the stigma of being a drifter.

So far, only selected data affecting the popularity of continental drift have been described. Having rescued the hypothesis from the banks of the Styx, where it had been consigned unfairly, the time has come to summarize some of the evidence for and against drift as science sees it today.

Biological and Paleontological Evidence (+ and −)

+ Strong correlations among the Permian flora of South America, Australia, India, and Antarctica. Shells, amphibia, and reptiles are also correlated. The discovery of the labyrinthdont fossil in Antarctica in 1967 greatly strengthened these associations. These amphibians had been found only in the South American and Australian rocks up to that time.

+ During the Devonian period a marine fauna with few elements in common with the fauna of the northern continents was present in South America, southern Africa, and Tasmania.

+ In 1967 the Canadian geologist J. Tuzo Wilson revealed that he had found Cambrian fossils in eastern Connecticut and Massachusetts similar to those in western Europe.

+ Lemurs, which are foxlike monkeys, are found today in India, Ceylon, Southeast Asia, Madagascar, and some parts of Africa. This evidence points to either a former land bridge across the Indian Ocean or continental drift. Of course, the popping up and down of land bridges or even the disappearance of a larger chunk of real estate (the geologists' lost continent of Lemuria) could explain many of the paleontological similarities between various continents. The single hypothesis of continental drift seems simpler.

+ There is some evidence that continental drift was the stimulus for the evolution of the more highly developed forms of bird migration; for example, some of the long overwater flights of land birds. Such a thesis is difficult to prove, but it accentuates the fact that continental drift can explain many diverse facts. This is an attribute of a good hypothesis.

− Many paleontologists feel that the correlations are not as strong as the drifters make them out to be. In his enthusiasm Wegener probably went overboard in this respect. For example, the early Tertiary faunas of eastern North America and western Europe are not nearly as closely related as one would expect if the continents were once fused. The same claim is made for the Devonian fauna of Brazil and southern Africa. Likewise, all of the terrestrial mammals of South America seem to have come from North America rather than Africa, suggesting a long continental separation from Africa. All of these objections are statistical in character; that is, everyone agrees that there are transoceanic similarities, but the argument hinges on just how similar faunas and floras have to be before drift is indicated. Of course the rates at which the fauna and flora diverged after continental fission occurred—a subject we know very little about—would strongly affect any correlations.

Geologic and Climatologic Evidence

+ Paleozoic and early Mesozoic sedimentary rocks in Brazil and Argentina are strikingly similar to those in southern Africa, India, and Australia. Some scientists place all of these rocks in the so-called Gondwana System.

+ Mountain ranges in western Europe seem to match ancient mountain chains in northeastern North America. To illustrate, the Great Glen fault in Scotland resembles the Cabot fault system which extends from Boston to Newfoundland. Also, the Sierras of Buenos Aires seem close cousins of the Cape Mountains at Africa's southern tip.

+ The jigsaw puzzle argument should be inserted here. Recent matching of continents along the central depths of their continental slopes have provided a much closer fit than many drifters had hoped for.

+ Akin to the jigsaw evidence is the recent finding by scientists from Massachusetts Institute of Technology and the University of San Paulo that the ages of rocks along the South American coasts and Africa correspond nicely where the puzzle pieces fit together. This *double* correlation of age and continental shelf outline is an extremely strong indication of drift. Further, there are sharp corresponding changes in age along the coasts, which makes the fit more convincing.

+ The distribution of Permo-Carboniferous glaciation strongly infers that South America, southern Africa, southern India, and Australia were all part of one continent during this period. Geologists can fit the continents together so that directions of ice travel (as indicated by scratches on rocks and the locations of glacial morraines) match.

+ The midoceanic rifts, the great grabens, the global cracks in the earth's shell, all indicate stretching forces arising from drift forces. Coastal mountain ranges can be explained as due to compression arising from the collision of crustal plates.

+ The ocean floors are definitely very young, indicating that drift might have occurred. For example, the Atlantic Ocean very likely did not even exist 150 million years ago.

+ Several island chains, such as the West Indies, fit the superficial description of debris left in the wakes of drifting continents. This is an intuitive feeling, though, and like the jigsaw puzzle argument may have no bearing at all on the reality of drift.

+ In addition to the glacial evidence, the fossil coral reefs in Alaska, the Alps, and Argentina, and the coal measures in northern Alaska and Antarctica suggest remarkable climatic changes in the past. Continental drift is not the only hypothesis that can explain tropics being at our present geographical poles; for example, the earth might have been so hot that the tropics were pushed clear to both polar regions.

− About all that opponents of drift can say in the face of the mass of circumstantial geologic evidence is that it is strange that the continents began to migrate so late in geologic history—during the last 5 percent of the planet's history to be more precise. This is indeed an interesting question because it infers either an earth that is becoming more active, perhaps through internal accumulation of radioactive heat, or an earth that saw great astronomical catastrophes very recently. Possibly, as the noted physicist P. A. M. Dirac has suggested, the gravitational constant that helps glue matter together has weakened during the last 150 million years.

Geophysical Evidence

+ Paleomagnetic measurements strongly support continental drift, although other interpretations are still proposed.
+ Magnetic anomaly data from the sea floor support both the sea-floor spreading theory and the internal convection-cell hypothesis, both of which mesh nicely with the idea of continental drift.
± Astronomical measurements of actual drift remain inconclusive at present.

In short, a great many different kinds of facts can be explained with the continental-drift hypothesis. We are also fortunate enough to have a more realistic physical cause of drift than Wegener ever had. Drift also has a certain subjective appeal absent in the concept of jack-in-the-box land bridges and the other conventional mechanisms accounting for the diverse facts. The concept of continental drift has a certain romance about it. A living dynamic earth has more appeal than a dead planet where all scientists can do is search endlessly through the strata for signs of a more exciting past.

The hypothesis of continental drift resembles Newton's Law of Gravitation in the sense that it gives science a single, conceptually simple way to account for a lot of different things. The doubters of drift should not be cast into the same category as those who plagued Galileo and Copernicus, although the language used and the desire for suppression were not dissimilar. In science the opposition party forces proponents of a theory to be careful with facts and not make exorbitant claims. The drift hypothesis still bears the marks of Wegener, Du Toit, and others, but it is a much stronger edifice than it would have been without its forty-year period of trial and ridicule.

Continental drift is once again a respectable working hypothesis for the earth scientist. Drift has not been proven, but the hypothesis seems much more reasonable than it did in 1960. It is now safe to say that the majority of the scientific community believes that the continents have been drifting apart during the past 150 to 300 million years, i.e., roughly since the beginning of the Cretaceous period.

More exciting than the fact of continental drift—assuming it is a fact, of course—is the implication that something has happened to the earth in relatively recent times. It is hard to resist the temptation to suppose that some cataclysm occurred in what is now the Pacific Ocean basin, and that this catastrophe cracked what remained of the original continental shell. The fragments are now drifting into the hole left by the cataclysm. On the other hand, continental drift may have been going on for much of the earth's history, and we now see only the latest rearrangement in process.

READING LIST

AMERICAN ASSOCIATION OF PETROLEUM GEOLOGISTS: *Theory of Continental Drift*, Tulsa, Okla., 1928 (Record of a historical confrontation of drifters and their opponents).

BAKER, H. B.: *The Atlantic Rift and Its Meaning*, mimeographed private edition, Detroit, 1932.

BARRETT, P. J., *et al.*: "Triassic Amphibian from Antarctica," *Science*, Aug. 2, 1968.

DU TOIT, A. L.: *Our Wandering Continents; An Hypothesis of Continental Drifting*, Hafner Publishing Co., New York, 1937.

GARLAND, G. D., ed.: *Continental Drift*, University of Toronto Press, Toronto, 1966.

HURLEY, P. M.: "The Confirmation of Continental Drift," *Scientific American*, Apr. 1968.

TAKEUCHI, H., *et al.*: *Debate About the Earth. Approach to Geophysics Through Analysis of Continental Drift*, Freeman, Cooper & Co., San Francisco, 1967.

WEGENER, A. L.: *The Origin of Continents and Oceans*, Dover Publications, Inc., New York, 1966 (paperback).

WILSON, J. T.: "Continental Drift," *Scientific American*, Apr. 1963.

5

UP-AND-DOWN
LAND BRIDGES

Ephemeral is the word for land bridges. Until a few years ago, whenever geologists and paleontologists were at a loss to explain the obvious transoceanic similarities of life that they deduced from the fossil record, they sharpened their pencils and sketched land bridges between appropriate continents. The eraser disposed of the bridge when it had outlived its usefulness as evidenced by the divergence of species on the sundered continents.

Is this cynicism justified? Perhaps, at least in the sense that opinion is now as firmly marshaled against ephemeral transoceanic land bridges as it once was against continental drift. (Of course, land bridges are really substitutes for continental drift.) We can laugh and look superior when we read the "old" geology texts of a decade or so ago as they describe several land bridges popping up and down between the Atlantic and Pacific shores, letting immigrants and emigrants dash across a narrow thread of still-wet sea bottom. And yet, in order to explain the manifest similarity, even identity, of species now separated by thousands of miles of ocean, some solid connections had to exist between the continents. At least, we suppose there had to be dry-land corridors. Insects, birds, and plant seeds can float, fly, and even be carried across ocean barriers on floating logs and in birds' entrails; life on earth has a remarkable wanderlust and colonizing ability. However, most reptiles, marsupials, and mammals find a water gap of a hundred miles impassable. One logical answer to the dilemma of species distribution was the land bridge. It was in fact the only acceptable answer during the 1920 to 1960 period when the concept of continental drift was interred (hopefully permanently) by the scientific community. Oh yes, there was a third possibility open: big sections of the earth's continental crust might have subsided beneath the waves Atlantis-style, breaking dry-land connections between continents. But Atlantis was legend, and the thought of really large tracts of foundering real estate was even more distasteful than continental drift during the first two-thirds of this century.

In analyzing the temper of geologic thought we must never forget that the doctrine of Uniformitarianism has held sway ever since James Hutton propounded it in the late eighteenth century. A Uniformitarian affirms that the geologic past must be explained in terms of processes we see transpiring today. And we do not physically see drifting continents, nor are any of the great land masses subsiding wholesale into the sea. However, we do perceive a slow inundation of local areas, such as the lands bordering the North Sea and the drowned river valleys of the Maine coast. Furthermore, no one denies that men and animals came and went across land that is now at the bottom of the North Sea and the English Channel; evidences of dry-land existence (stone tools and mammoth teeth) are dredged up regularly from these shallow arms of the sea. North Sea fishermen occasionally snag their nets on drowned forests. The Isthmus of Panama between two continents is an active land bridge that would disappear if the sea rose only a hundred feet or so. If the sea would subside or the land rise only five hundred feet, the famous Bering Strait land bridge would reappear just as it must have fifteen or twenty thousand years ago when man and other Asian species supposedly filtered across it down into North America. Thus, short land bridges are real today, and we can almost see them rising and subsiding. They are consistent with the principle of Uniformitarianism, while drifting and sinking continents strain the limits set on the motion of land masses by Uniformitarianism, which has been the most salutary, all-embracing principle ever stated in geology. When these facts are considered, one understands the tenacity with which geologists have clung to the land-bridge idea.

The concept of long transoceanic land bridges originated in the last half of the nineteenth century; the same time span that saw the first flowerings of many geologic hypotheses: Antonio Snider's continental drift, Louis Agassiz' Ice Ages, etc. The basic idea is usually attributed to the Frenchman Jules Marcou, who offered the land-bridge hypothesis in 1860 in his *Lettres sur les Roches du Jura*. The bridge concept was more firmly established in 1887 by Melchior Neumayr in his book *Erdgeschichte* (Earth history). Eduard Suess put land bridges in a global setting during the 1880's. His maps showed not only the unsplit southern supercontinent of Gondwanaland but also a North American peninsula that reached over to Europe via Greenland and Iceland. He called this thoroughfare for Cenozoic mammals *Atlantis*, but it did not last long enough to be the Atlantis of legend. After his Gondwanaland broke up during the age of dinosaurs, Suess believed that a land bridge connecting South Africa with peninsular India remained to provide interchange of species between the two lands. The biological correlations between India and Madagascar are so strong that some sort of land connection in the geologic past was essential. The great Seychelles reefs, with their cores of granitic continental rock lie between India and Africa. Geologists figure they must be remnants either of: (1) a now-sunken land bridge; (2) the foundered continent of Lemuria; or (3) debris left in the wake of India's pilgrimage across the ocean to where it now lies against the Himalayas (which drifters claim were wrinkled into existence by the force of the collision).

Carried away by the work of Suess and Neumayr, geologists like John Gregory filled the oceans with networks of land bridges. Wherever paleontology seemed to require a sunken continent or continental drift to explain correlations of fauna and flora, a land bridge would come to the rescue. Of course, the short Panamanian and Bering land bridges are uncontested, but there is no doubt that the bridgers went too far in terms of the number of bridges they manufactured and their lengths.

Three big bridges spanned the Atlantic:

Archiboreis stretched from North America, across the present arctic islands (like Suess's Atlantis bridge) to northern Europe.

Archatlantis went along the West Indies chain of islands to North Africa.

Archhelenis connected Brazil with southern Africa.

In the Pacific the bridges had to be even longer if stationary continents are assumed:

Archigalenis ran from central America, through the Hawaiian chain, to northeast Asia.
Archinotis provided the desired route from South America to Antarctica.

Many geologists opposed building land bridges with such wild abandon. None could deny, however, the convincing evidence supporting the Bering, Panamanian, and Lemurian land bridges. Archiboreis did not seem too farfetched either. Somehow, though, the big systems of transoceanic bridges seemed too facile, too artificial.

During its heyday the land-bridge concept was continually refined by such scientists as George Gaylord Simpson, an ardent bridger and outspoken antidrifter. Simpson believed that many land bridges existed during our geologic past, so many that he classified them into three categories: (1) corridors, (2) filter bridges, and (3) sweepstakes bridges.

Corridors are defined as broad, continuous land connections existing for long periods of time, such as that that once existed between Europe and the British Isles. There would be extensive exchange of fauna and flora between the two continents. Suess's supercontinent, Gondwanaland, for example, could be supplanted by suitable corridors between permanently fixed southern continents.

A filter bridge is narrower than a corridor and need not be completely continuous. It sees the sun for a shorter period of time before subsiding into the sea. Because of their geographical narrowness, filter bridges offer potential emigrants much more limited climatic and ecological environments. A much narrower spectrum of fauna and flora will be attracted across a filter bridge; in other words, the bridge environment acts as a biological filter. The Bering land bridge was of the filter type, so was Archiboreis, which supposedly connected North America and Europe. Because of the filtering, arctic fauna and flora are remarkably similar around the world; they passed through the filter, while few temperate zone plants and creatures cared to make the trek across the frigid bridge. Bears, cats, bison, deer, and mammoths crossed eastward into North America, while dogs, horses, and camels seem to have gone in the other direction.

Water barriers or perhaps high mountain ranges exist along

Simpson's sweepstakes routes. Interchange of species across these bridges is improbable; in Simpson's analogy only a few animals hold winning tickets. If it is a water barrier, like that between Africa proper and Madagascar, only a few small animals clinging to driftwood and uprooted trees make the crossing. Island life in the Pacific was apparently established via sweepstakes routes. Island fauna and flora are likely to be unbalanced because of the extreme filtering action. Sweepstakes routes are conceptually useful because they extend the land-bridge idea to chains of islands. No longer is the geologist hard put to provide long continuous threads of dry land between continents.

To expand on the geologist's problem, how could thousand-mile-long dry-land bridges be built and then torn down? The bridgers had always scoffed that continental drift was impossible because there seemed to be no natural forces capable of moving whole continents. When pressed for bridge-building mechanisms, however, the bridgers could do little better than the drifters. The Bering and Panamanian bridges were no trouble because everyone admitted that drastic changes in sea level had occurred in the past, particularly during the Ice Ages, when so much of the earth's water inventory was frozen into the great mile-or-more-thick ice sheets. As for the transoceanic links, the bridgers maintained that they must have been created by ocean-bottom mountain-building processes (orogeny) such as those at work on the continents. Wrinkling forces arising from the cooling, contracting earth would squeeze up a fold of crust several thousand miles long (a geoanticline); ultimately this would subside, severing the land bridge. Were there not many long continental mountain chains to show how it was all done?

Exploration of the sea bottom has failed to find any strong evidence that long, transoceanic geoanticlines ever broke through the sea's surface. Instead, the ocean-bottom ridges seem to bisect ocean basins between land masses rather than bridge them, although some underwater ridges do come ashore—the East Pacific Rise does at Baja California. Furthermore, these underwater ridges show no signs of ever having been exposed to the open air, although there are a very few, rather puzzling reports of sand and shallow-water deposits being dredged up from the mid-Atlantic Ridge.

However, ocean-bottom sounding and sampling in the Pacific

did reveal something to hearten the geologists and paleontologists looking for an alternative to continental drift: the *guyots*. The guyots are flat-topped, wave-planed seamounts (also called tablemounts) found in profusion in the Pacific and Indian oceans, and to a lesser extent in the Atlantic. Named after the Swiss-American geologist, Arnold Guyot, thousands of guyots are sprinkled around the Pacific basin, many in linear chains. The flat tops of the guyots are not shallow-water features in today's oceans; they are usually located over a thousand feet below the surface, with a great many being a full mile and more under the waves. Neither are they small features—the Great Meteor Seamount in the Atlantic has a flat top the size of Rhode Island. The bridgers seized upon guyot chains as obvious remnants of old land bridges or at the very least transoceanic stepping stones.

What perplexes oceanographers most are the samples they dredge up from the guyot tops: nicely rounded and obviously abraded (probably water-worked) volcanic debris and samples of coral from the Upper Cretaceous period (about 100 million years old; the oldest fossils retrieved from any ocean). Coral samples that could not have grown in water more than five hundred feet deep have been retrieved from guyot tops a mile down.

Present thinking states that the guyots are basically volcanic in origin, but that they once poked their cinder cones above the surface near active ocean rifts and were planed flat by wave action. As the sea floor spread away from the rift, the guyots were carried along. After 20 to 30 million years the crust sank to the point where the flat-topped guyots are thousands of feet below the surface. The farther the guyots are away from the rift the older and deeper they are. For example, the guyots are found almost exclusively on crust that is more than 30 million years old. This theory of crustal plate subsidence as it moves away from the rifts explains both the flat tops of the guyots and their tendency to occur in chains parallel to the ocean rifts. An alternative hypothesis involves a large-scale drop in sea level within the past 100 million years. A drop in sea level would explain the flat tops but not the linear arrangement of many guyots.

In the context of land bridges the guyots potentially offer long chains of stepping stones across the Pacific during the early Tertiary and Cretaceous periods. Insects, birds, and perhaps some land animals might have crept stepwise across the Pacific. Short reaches of open water are not significant barriers to many forms of life; guyot stepping stones might have substituted for continuous land bridges.

All this about guyot-hopping animal traffic is still surmise, of course. If oceanographers could bring up some fossil bones of animals caught by death en route between continents, and if a reasonable explanation of such tremendous changes in sea level could be offered, this discontinuous type of land bridge would be more believable.

There is plentiful evidence, however, that sea levels have changed a few hundred feet in recent times. Let us look in detail at the Bering and Lemurian land bridges: the first is a well-established bridge, the second a substitute for the long drift of peninsular India from South Africa northeastward to where it now abuts mainland Asia.

For those who still believe in the permanency of the continents and ocean basins, the Bering Land Bridge is an article of faith. (Of course, one can believe in *both* land bridges and drift and still be consistent. They are not naturally exclusive.) The Bering Land Bridge is the showcase of the bridgers; the land bridge no one seriously questions. Nature has exposed and covered this biological conduit several times in recent geological time. We see the effects in terms of the distribution and evolution of animals, particularly mammals and man himself, and also in the interchange of marine life between the Pacific and the Arctic Oceans. A land bridge is an ocean barrier when it is down and a land barrier when it is up. The Bering Land Bridge has worked just the way a filter bridge should; east and west when it was dry, north and south when it was wet.°

When the Spanish were exploring the New World some four centuries ago, the Asiatic cast of the American Indians was quickly noted. In 1590 Fray José de Acosta postulated that a narrow strait or land connection must have existed in northern climes over which small bands of hunters first entered North America. Actually, the aborigines in North and South America are surprisingly diversified from the anthropologist's standpoint. Bridgers have had to postulate several waves of immigrants entering via the Bering

° A modern example of a sweepstakes water route is the Panama Canal, through which a number of Atlantic and Pacific marine species have been swapped.

Land Bridge to account for the varied Indian cultures. Even then there seem to be some anthropological loose ends; and a small contingent of scientists believes that some South American immigrants may have sailed directly, though perhaps inadvertently, across the Pacific from Asia. And a still smaller group (boasting few scientists) maintains that it sees evidence of colonization from lost Atlantis in the culture and physical characteristics of Indian races around the Atlantic basin. Thor Heyerdahl's 1969 voyage in his papyrus boat, the *Ra*, was an attempt to show how transatlantic contacts from Egypt could have been made in ancient times. Despite this discordance in the background the rest of the orchestra has consistently played the Bering Land Bridge theme ever since Fray José de Acosta.

When Vitus Bering sailed through the strait bearing his name in August 1728, he proved once and for all that North America and Asia were physically separated. Had it not been a foggy day, Bering could have seen both continents, for they are separated by only 55 miles. Had he been using the ship's sounding line, he would also have found that he was sailing over a wide plain that lay only a few hundred feet below his keel. In the strait itself the maximum depth is about 170 feet. Here also the two Diomede Islands and Fairway Rock, the last remnants of the land bridge, jut out of the cold water, hinting of the land connection just below the surface. Indeed, between October and June the Strait is frozen over, and a hardy man could walk from Russia to America.

During each of the several ice ages that the world has seen during the past few million years, the Bering Land Bridge became exposed as the continental ice caps began to grow, accumulating water that came ultimately from the sea. The geologist Eric Hulten christened the broad arctic lowland exposed by the receding sea *Beringia*. Rather than envisaging wolves, mammoths, and other arctic creatures as trooping west to east across the bridge, it is probably more accurate to think of Beringia and its environs as low-lying refuge for animals being driven in front of the advancing American and Asian ice sheets. After the retreat of the ice the animals dispersed in all directions.

The number of ice ages in the Bering region varies with the investigator, but ten seems to be the minimum. (Only four or five distinct glaciations are recognized in middle North America.) Beginning two or three million years ago, ice caps advanced and retreated on the Northern continents, exposing the Bering Land Bridge with each advance. Supposedly, a new wave of immigrants, including man, spread into North America with each retreat of each ice sheet. Camels and a few other American species were introduced to the Old World at the same time, but most traffic came east. During the interglacial periods—today, for example—we also see the exchange of marine life between oceans. There is little doubt that an ephemeral land bridge repeatedly linked Asia with North America. But the reality of the fifty-five-mile Bering Land Bridge does not prove the past existence of transoceanic land bridges any more than Arizona's Meteor Crater proves that the Pacific basin was blasted out by a planetoid.

Do we have any evidence supporting the past existence of *long* land bridges? The best place to look is in the Indian Ocean between southern Africa and peninsular India. Here the mid-Indian Ridge curves northeastward from below the Cape of Good Hope toward Ceylon and the southern tip of India. Before it reaches these targets, it abruptly ends as it hits another section of the mid-Indian Ridge running southeast from the Red Sea and the African Rift Valleys. Between continental Africa and these sections of the world-encircling ridge-rift system lie Madagascar, the Seychelles, the Chagos Archipelago, the Maldive Islands, and many other fragments of land. The Seychelles, Madagascar, and some of the other islands are not volcanic but continental in shape and composition. Because of their small sizes, they have been dubbed *microcontinents*. In spots the sea bottom is under more than a mile of water, but there does seem to be the backbone of a bridgelike structure made of continental rock. Many guyots are also found in the Indian Ocean, and it is just possible that a dry-land bridge extended from Madagascar to India's tip during the period when sea level was very low and the guyots were being planed flat by wave action.

The paleobiological evidence for an Indian Ocean land bridge is very impressive. In fact, biologists often say that Madagascar's fauna is much more like that of India than mainland Africa, just 300 miles to the west. There are also similarities between African animals and those in India and Southeast Asia. Historically, the most important animal that helped build the land bridge idea is the lemur, a primitive primate. Lemurs were first found in Africa, Madagascar, and Southeast Asia, strongly inferring a recent land connection between these lands. There are many similar biological connections between these continents and microcontinents. A dry land connection from Cambrian times to almost the end of the Cretaceous period (75 million years ago) seems a simple solution to these distribution problems.

The same sort of evidence indicates that the dry land connection between the African mainland and Madagascar disappeared much earlier, probably near the end of the Triassic period (180 million years ago), the same time that the supercontinent Gondwanaland broke up, according to the drifters. However, the Madagascar-India land link seems to have persisted until that mysterious, probably violent, time, some 75 million years ago, when the earth saw many old things end and new ones begin.

The furry, long-tailed lemurs provide the name for the long land connection between Madagascar and India—Lemuria. But was this land connection really a bridge? The facts we know today can be interpreted in three different ways:

(1) A long, now-inundated land bridge extended from Madagascar to India until the end of the Cretaceous period.
(2) Madagascar, the Seychelles, and a few other Indian Ocean islands, with their granitic cores, are actually the remnants of the fabled lost continent of Lemuria (or Mu), which supposedly foundered near the end of the Cretaceous. The difference between a continent and land bridge is primarily one of size.
(3) Peninsular India and Madagascar were once united but split at the end of the Cretaceous, when India drifted to its present position, nudging Asia and creating the Himalayas during the collision. In its wake, India lost the Seychelles and sundry continental debris.

The logical equivalence of long land bridges, sunken continents, and continental drift is apparent in the problem of Lemuria. Each of the three mechanisms explains many of the facts.

The history of the Lemurian land bridge has important connotations for subsequent chapters. The first inklings of some sort of transoceanic connection between Africa and India came during the 1860's and 1870's, when geologists, such as William T. Blanford, pointed out the striking resemblances between certain Permian rock formations in India and South Africa. The Indian rocks made up what was called the Gondwana formation, which eventually lent its name to Suess's Gondwanaland.° Blanford and his contem-

° Gondwana means "land of the Gonds." The Gonds were a forest tribe in India that tortured people to death to make their crops grow.

poraries assumed that a land bridge once existed between the two continents. It was the English zoologist Philip L. Sclater who explained the unusual distribution of lemurs with a land bridge called Lemuria.

During the same period, Ernst Heinrich Haeckel, the leading proponent of Darwinian evolution in Germany, suggested that the Africa-India land bridge might have lasted clear into the Cenozoic Era, the Age of Mammals. Perhaps, he went on, this sunken land was the original home of man. In the front of his popular 1868 book, *History of Creation,* Haeckel placed a map of Lemuria, making the lost continent seem even more real. Other scientists went on to state that the sinking of Lemuria was only the most recent event in the breakup and dispersal of the supercontinent Gondwanaland. Eventually, however, lemur fossils were discovered in Europe and North America and scientists saw no real need for a recent land connection between India and Africa. Lemur distribution no longer needed Lemuria. Severance of India and Africa was pushed back from the Age of Mammals to the Cretaceous period (part of the Mesozoic Era), although it might have persisted a little longer.

Before the scientists decided they could do without Lemuria, the occultists took possession of this lost land. Indeed, the occultists of today foresee imminent terrestrial catastrophes—such as that which did away with Lemuria—that will submerge and cleanse much of the earth, particularly the Los Angeles area.

The later occultists have enraged scientists with their clairvoyant musings. Detailed maps and city plans of Lemuria and Atlantis have appeared out of the ether or whatever medium conveys such information. Our Lemurian ancestors have been described at great length. Parapsychology may have achieved some minimum scientific stature in recent years, but organized science will have nothing to do with the occultists and their visions of Lemuria. The consequence has been scientific backlash; foundered continents are now dismissed out of hand. Scientists have had their fill of Lemuria and Atlantis. Continental drift is infinitely more acceptable in today's intellectual climate.

To summarize, short land bridges exist today and certainly did in the geologic past. The evidence of paleontology and the present distribution of fauna and flora strongly suggest that the continents were once linked either by long transoceanic land bridges or by actual continental contact prior to drift. There is really not enough information to specify "bridge" or "drift" in every specific case. Geologists tend to favor bridges, while geophysicists (since 1960, at least) have become ardent drifters. The major points of continental contact may be summarized as follows:

Connection	Current Opinion
Asia-North America	Uncontested land bridge.
North America-South America . .	Uncontested land bridge.
North America-Europe	Broad bridge severed by drift.
Africa-Madagascar-India.	Continental contact followed by drift slightly favored.
Africa-South America and Australia-Antarctica	Continental contact followed by drift. (Breakup of a supercontinent is equivalent to multiple bridges.)

READING LIST

DE CAMP, L. S.: *Lost Continents,* The Gnome Press, Hicksville, N.Y., 1954.

HOPKINS, D. M., ed.: *The Bering Land Bridge,* Stanford University Press, Stanford, Calif., 1967.

TAKEUCHI, H., *et al.: Debate About the Earth. Approach to Geophysics Through Analysis of Continental Drift,* Freeman, Cooper & Co., San Francisco, 1967.

WEGENER, A.: *The Origin of Continents and Oceans,* Dover Publications, Inc., New York, 1966 (paperback).

6

GRAND CANYONS
BENEATH THE SEA

Strip away a mile or so of ocean water—roughly half the world's inventory—and the tourists would eschew the Grand Canyon of the Colorado for the greater vistas of the continental fringes. With the continental blocks towering high above shrunken seas, one could see great canyons incised in the sharp rims of continental shelves; incisions that the eye could trace for tens, even hundreds, of miles seaward down the continental slopes. It would be eerie standing on the lip of a continent; behind would be miles of desolate mud and sand where fishermen once trawled shallow waters; ahead the land's rivers would rush through the deep gorges toward a distant sea lapping at the foot of the massive continental crustal block. A few land bridges would tie some of the continents together, and thousands of flat-topped guyots would rise above the ocean. The great ocean trenches, however, where one crustal plate disappears under another, would still lie far beneath the shrunken sea's surface. Could such a sea-poor world ever have existed in the past? The answer may come soon as dozens of scientists explore the submarine canyons in small research submarines.

Although the submarine canyons cannot be seen with the eyes, they were too big to escape even the crude sounding lines in use a century or two ago. As mariners sounded the approaches to the world's seaports, huge drowned valleys began to appear on their navigation charts. The noted American geologist James D. Dana was apparently the first to call attention to the undersea continuation of the Hudson River estuary in his 1863 classic, *A Manual of Geology*. Based on extensive soundings made as far as sixty miles out at sea by the U.S. Coast and Geodetic Survey, A. Lindenkohl in 1887 again called attention to this remarkable gorge with walls some four thousand feet high in places. (Few Manhattanites dream of what lies at their seaside doorstep.) During the late 1880's fairly detailed maps were made of several submarine canyons. As they were traced seaward with the sounding line, the canyons looked very much like the typical river-cut canyons seen on dry land. It was only logical that most geologists of this period classified them as true river-cut gorges that had been drowned as the land fell or the sea rose in past aeons. Other mechanisms, such as ocean-floor currents, were suggested, but the drowned-river idea sounded good. Were there not obvious drowned rivers and bays around the shores of the North Atlantic?

Around the turn of the century, however, A. C. Spencer focused critical attention on the canyons when he proposed that they were cut during great continental uplifts, which, he said, also caused the Pleistocene glaciations. His theory tied in with the hypothesis that sea levels were a lot lower during the Ice Ages because so much of our planet's water inventory was tied up in the ice sheets. Spencer proposed elevating continents, while others hypothesized lowered sea levels. Both processes would have given terrestrial rivers an opportunity to cut away at the edges of the continental shelves. Perhaps both processes worked together. The difficulty was that the submarine canyons, if the soundings were correct, called for sea-level changes of many thousands of feet within the past fifty thousand years or so. Such violent changes could not be countenanced under the prevailing philosophy of Uniformitarianism. To quote Shepard and Dill, in their *Submarine Canyons:* "It smacked of the outmoded ideas of catastrophism." Rather than reconsider Uniformitarianism, the very foundation of geology, geologists preferred to forget about submarine canyons. For a quarter of a century, submarine canyons languished unseen not only by the ships that sailed over them but by geologists in general. Some even suggested that they did not really exist. Quoting Shepard and Dill again: "When an observation offends one's preconceived ideas, it is easy to dismiss it."

When the echo sounder came into use in the 1920's, the submarine canyons were once more revealed in all their magnificence and mystery. They could not be ignored any longer. The thirties saw an amazing variety of canyon-cutting mechanisms proposed, all essentially Uniformitarian in character. As we shall see, many of these hypotheses are still around, even though modern oceanography with its coring rigs, research submersibles, and greater financial resources has sharpened the scientific picture a great deal. Before accepting or discarding any hypotheses, a close look at the canyons is in order.

The submarine canyons typically display V-shaped profiles with high steep walls. Like terrestrial river gorges they cut their ways through all types of rocks, taking the paths of least resistance. Numerous tributaries join the main canyons on both sides. Some trib-

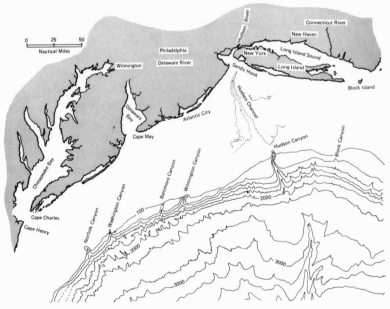

Map of some of the submarine canyons off the East Coast of the United States. All cut deeply into the lip of the continental shelf. The famous Hudson Canyon is the only one obviously associated with a terrestrial river. Note also that the Hudson Canyon can be traced far down the continental slope. (Adapted from F. P. Shepard and R. F. Dill, *Submarine Canyons and Other Sea Valleys*, p. 146)

utaries intersect the main canyons high up on the walls after the fashion of the "overhanging" tributaries seen in the canyonlands of the American West. A contour hydrographic map of a submarine canyon strongly resembles a topographic map of a terrestrial canyon; and one is sorely tempted to agree with the scientists of almost a century ago when they hypothesized a river-cutting mechanism.

Of course there are other gullies, gorges, and gashes beneath the sea. The great rift that tends to bisect the midoceanic ridge system is a canyon of sorts, but its origin seems tectonic, as do the deep ocean trenches discussed in chapter 3. A few offshore valleys, such as the San Clemente Rift Valley off southern California, are also probably grabens or rifts of tectonic origin. Some U-shaped or troughlike valleys that cut into the continental shelves were probably gouged out by glaciers, for these occur frequently off glaciated coasts. Beyond the submerged mouths of the true submarine canyons lie the so-called *fan-valleys* (Shepard-Dill terminology). Fan-valleys have relatively low walls and often have raised rims that are analogous to the natural levees found along river channels, such as the Mississippi, that cross deltas. Sea-valley classification is far from clear-cut, and some valleys are doubtless hybrids. Of all the various sea valleys, it is the V-shaped submarine canyon that has cut its channel deep into the continental rocks that attracts our attention because it is a feature that may provide clues about the earth's past.

The true submarine canyons begin at the edges of all continents as well as outlying islands, such as the Bahamas and the Hawaiian group. Many, though far from all, seem to be associated with extant terrestrial rivers—the Hudson Canyon, for example—or with terrestrial rivers that have been diverted into new channels in relatively recent geologic time. Then there are other canyons engraved in the continental shelves without the least hint of a shore-based origin. These confound those who propose that the canyons were river-cut when the continents were high above the sea.

Discounting the terminal fan valleys, the Bering Canyon in the southern Bering Sea is the longest—230 miles. The average canyon runs only for about 35 miles. If the fan-valleys are included, some of the bigger canyons stretch out for 500 miles. The canyon heads may begin almost at the present beach or as deep as a thousand feet below the water. The deepest canyons end at depths of about

15,000 feet below sea level. The heights of the canyon walls vary considerably, but no one can deny that the biggest are the deepest gashes in the planet's crust, on land or sea bottom. The Great Bahama Canyon, for example, has walls three miles high (three times as high as the walls of the Grand Canyon). Shepard and Dill give an average maximum wall height of about 3,000 feet for the limited number of canyons they knew about. At least three other submarine canyons outclass Grand Canyon.

The canyons definitely have currents in them, even though they are submerged under many feet of water. The currents are rather slow, generally less than a mile per hour, and the directions of flow sometimes reverse. Cores taken from the canyon bottoms yield mud and sand, indicating that the currents tend to sweep sediment down the canyons toward the fan-valleys. Without these currents the canyons would quickly fill. By scraping the canyon sides with sampling equipment, pieces of sedimentary and crystalline igneous rock, mostly granite in the latter instance, have been brought to the surface.

The age of the submarine canyons has not been measured; perhaps *ages* would be more appropriate because they may have originated at different times. The canyons obviously have to be younger than the youngest sedimentary rocks they cut through. Some scientists believe that all of the submarine canyons were cut during and after the Ice Ages, which would make them only a few tens of thousands of years old. Shepard and Dill, however, feel that the canyons may be relatively ancient features of the earth that have been slowly etched away, aeon after aeon, sometimes under a canopy of water, sometimes in the open air. Unfortunately, there is no precise way to date a canyon.

A brief vignette of a representative submarine canyon fills out the physical picture. The Hudson Canyon on the East Coast and the several canyons off southern California and Baja California have been explored and described in greatest detail, mainly because they are conveniently located with respect to Lamont Geological Observatory, Woods Hole Oceanographic Institution, Scripps Institution of Oceanography, and other research groups. But rather than select a well-known canyon, let us examine the less-publicized Great Bahama Canyon; first, because it is one of the biggest canyons and, second, because the Bahamas will reappear in the next chapter.

J. W. Spencer first noted the existence of deep submarine valleys in the vicinity of the Bahamas in his 1895 paper: "Reconstruction of the Antillean Continent." In the 1930's and 1940's some surveys with echo sounders were made, but it was not until the 1960's that modern echo sounding and navigational equipment was applied in the area by the Woods Hole Oceanographic Institution. This survey established that a deep, V-shaped canyon runs for more than forty miles through the center of a great embayment or shallow depression called the Tongue of the Ocean. This general area has since become a U.S. Navy proving ground for submarine and anti-submarine techniques.

The Tongue of the Ocean is a deep, bowl-shaped basin, some 20 miles wide, that seems scooped out of the otherwise shallow shelf just east of Andros Island. The sides of the Tongue drop off rapidly to a rather flat floor more than a half mile deep. Northward, however, the centerline of the Tongue plummets to form the head of the Great Bahama Canyon. The main canyon swings northeast around New Providence Island, cutting through the continental shelf to a depth of at least 14,000 feet. It is at this point that the V-shaped walls rise nearly three miles above the canyon floor. The canyon continues seaward, changing into two trough-shaped valleys at a depth of about 15,000 feet. The entire canyon is approximately 125 miles long. In addition to the main canyon, the Northwest Providence Branch of the canyon strikes northwestward from the Tongue of the Ocean at a depth of about 12,000 feet. This canyon is not as impressive. Both branches are joined by numerous tributaries. In passing, it should be noted that several other canyons, smaller in size but much steeper, have been discovered around the Bahamas.

Photographs taken at a depth of 11,500 feet in the Great Bahama Canyon reveal rocky walls hemming a sandy floor crossed by ripples and dotted with rounded cobbles. Both ripples and exposed cobbles hint at the existence of currents in the canyons.

There are no existing terrestrial rivers that could have scoured out the Tongue of the Ocean and the great canyons that open up north of it. Little wonder that geologists have debated spiritedly and sometimes vitriolically among themselves for decades about these deep gashes appearing along the continental shelves and on the flanks of midocean islands.

Although the submarine canyons look superficially like terrestrial river-cut canyons, modern geologists usually avoid this hypothesis like the plague because of its Catastrophic overtones. In fact, they have proposed almost every other mechanism that Uniformitarianism permits. The first three of these ideas have been pretty well discounted by modern discoveries, but they illustrate the rather frantic search that has been made for reasonable hypotheses.

In 1893, when the outlines of the canyons first appeared on bathymetric charts, Andrew Lawson ascribed them to faulting. Wegener did also in 1924. Although a few submarine valleys are doubtless fault troughs or grabens, many long sinuous submarine canyons extend seaward, oblivious to the general structural trends of the continents and cannot be fault-created. Spring sapping was suggested by Douglas Johnson in 1939. In this view water under pressure in rock layers extending out under the continental shelves would boil out of the rock (as in an artesian well) where the rock formation ended on the continental slope. This subsurface current would erode a channel down the side of the continental block. Seismologists now assure us that few if any water-laden rock layers (aquifers) are associated with submarine canyons. In 1940 Walter A. Bucher thought that seismic sea waves (tsunamis) might slosh against the continental shelves, creating deep subsurface currents that might scour out the canyons over the millennia. Field studies and the frequent association of the canyons with terrestrial rivers have relegated this idea to the discard pile.

Two mechanisms still in contention are submarine landslides and subsurface currents. Landslides are known to occur near the heads of some canyons. However, according to Shepard and Dill, they seem sufficient to explain only some erosion near the heads. Of course, slumping and other mass movement also occur at the deep-sea end of the canyons where sediment accumulates. The major criticisms leveled against the landslide idea are: (1) the submarine canyons are more branching (dendritic) than typical terrestrial landslide canyons, and (2) we do not have evidence of landslides in the main portions of the canyons.

It has also been proposed that the ordinary bottom currents might have cut the canyons over long periods of time. These currents are generally very weak—less than one knot—and may move upcanyon as well as down; they also seem to correlate with the tides. Conceivably, these unimpressive currents might erode a canyon miles deep, given enough time, but they certainly do not have the cutting power of the Colorado. It is also hard to imagine why supposedly random subsurface currents would start to gnaw canyons out of shelf rock at the same spots where so many terrestrial rivers empty into the sea.

Summarizing, landslides and bottom currents seem to be rather weak hypotheses.

Most geologists now believe that the submarine canyons were eroded by *turbidity currents*—streams of fluidized mud and sand that sometimes cascade down the continental slopes, perhaps in response to an earth tremor. In action, turbidity currents resemble avalanches except that they are not flows of relatively solid sediment. Rather, the mud and sand in turbidity currents are intimately mixed with water and held in suspension as the whole mass moves downslope under the pull of gravity. The density of a turbidity current is much lower than the consolidated sediment in a landslide but significantly higher than the water in canyon bottom currents. Actually, turbidity currents combine the fluidity of water with the speed of the landslide.

Reginald A. Daly proposed turbidity currents (he called them *density currents*) as a causative factor in canyon formation in 1936. There seems little doubt that turbidity currents frequently sweep down the continental shelves, cutting through soft sediments. Furthermore, laboratory experiments have shown that dense flows of fluidized sand and mud have awesome cutting power when churning down slopes at ten or twenty knots. To some extent turbidity currents probably help excavate the upper and lower portions of submarine canyons, just as landslides do. Geophysicists, such as Bruce C. Heezen, have suggested that the submarine canyons were cut by turbidity currents during the Ice Ages when the seas were several hundred feet lower than they are now when mainland streams run clear out to the outer rims of the continental shelves. As the streams flowed over the shelf lip they fluidized the sediments on the slopes, giving birth to turbidity currents that roared down the slope like liquid sandpaper. The turbidity current hypothesis has much to offer and is manifestly within the domain of a Uniformitarian.

Those who believe strongly in turbidity currents as erosive agents point to the aftereffects of the Grand Banks earthquake of November 18, 1929. Many transatlantic cables are laid across this region. Some of these parted during the first sharp shocks. However, other cables hundreds of miles downslope did not break for hours afterward. If the cables' distances from the earthquake epicenter are plotted against the times at which cable service ceased, a straight line results. The interpretation usually given is that the earthquake loosed a colossal mass of fluidized sediment, which swept downslope along a front over a hundred miles wide with an initial speed of some fifty-five knots, slowing to about fifteen knots on the flatter slopes. Cores taken in the area afterward showed old layers of undisturbed sediment, thus tending to refute the concept of an immense rush of liquefied sediment. Instead of a massive wall of fluidized material, localized turbidity currents triggered by aftershocks could have caused the same pattern of cable breaks. In short, no one really knows exactly what happened in 1929. Much more work will have to be done with coring rigs and submersibles before the complete story is known.

Although some question exists about the role of turbidity currents in the Grand Banks incident, no one denies their existence. Shepard and Dill argue against assuming that turbidity currents are the only or even the major eroding agents in the submarine canyons. To begin the case for the prosecution, the fine, gently rippled sands and silty clays found in the bottoms of the canyons today imply only weak currents, not torrents of fluidized sand and mud that must act like purgatives in the narrow canyon bottoms. Bathyscaphe dives after seismic disturbances off the California coast found no turbidity currents were thus stimulated. All evidence is that present-day currents along most of the canyon floors are too weak to have excavated such tremendous volumes of rock in a few tens of thousands or even several millions of years.

Two more subtle and intuitive difficulties are raised by the turbidity-current hypothesis. First, many canyons are winding and tortuous, some have rock overhangs. Considering the vaunted violence of the turbidity current, one would expect relatively straight and direct thoroughfares to be scoured out of the rocks of the continental shelves. There should be few tributaries. Instead, the canyons show the filigree work and number of tributaries usually found with water sculpture in terrestrial rivers. (The same argument was used against the landslide hypothesis.) The second subtle objection is that some canyons, notably the Great Bahama Canyon, head in deep water where there is little source of sediment to feed turbidity currents and where no major terrestrial river exists to generate turbidity currents.

Despite the objections, turbidity currents are generally felt to be the most logical causative agent in canyon formation. The author, however, has a purpose in promoting theories associated with that discredited philosophy called Catastrophism. For this reason, the river-cutting hypothesis has been saved for last.

If we could siphon off ten thousand feet of ocean water, few terrestrial geologists would hesitate to designate the newly exposed submarine canyons as bona fide river canyons. *They look like river canyons.* Not a very profound reason in a world where so many features of nature are not what they seem, but we can go a few steps further because some submarine canyons—notably the Tokyo Canyon and some around Corsica—are clearly seaward continuations of land canyons. Even the most enthusiastic proponents of turbidity currents grant that some submarine canyons are drowned river valleys. But are they all?

Put this question to most geologists and they will reply in the negative and emphatically at that. River erosion of the canyons would require the oceans to be six to ten thousand feet below the lips of the continental shelves; this, most cannot countenance. An exception was A. C. Veatch, who in the October 1941 issue of *Scientific American* suggested that sea level was some twelve thousand feet below its present level during the Ice Ages. It was during this period that the submarine canyons were cut and, according to Veatch, Atlantis thrived on the exposed mid-Atlantic Ridge. Evidence for the existence of fabled Atlantis is scant, but one does find some persuasive geologic evidence for lower sea levels a hundred million years or so ago. For example, wells drilled along the coasts commonly encounter thousands of feet of terrestrial or shallow water formations dating back to the Cretaceous period (when continental drift probably began). In addition, deep borings in the Bahamas reveal nothing but shallow water formations, indicating that the Great Bahama Canyon may have been drowned quite recently. These are at best only innuendos of Catastrophism.

The river-erosion hypothesis would be more believable if there were some reasonable mechanism available to either raise the sea level a couple miles or depress the continental blocks the same amount. The great ice sheets covering the northern polar regions during the Ice Ages could have impounded enough water in the form of ice to drop the sea level by the few hundred feet needed to expose the Bering Land Bridge, but ten thousand feet? The ice sheets would have had to be ten or fifteen miles thick to collect that much water. The mile-thick ice sheets described in high school textbooks are hard enough to believe. Widespread glaciation a few millennia back cannot be questioned, but a ten-mile-thick ice sheet grates against the intuition.

Others have gone further down the Catastrophic route, proposing that the great changes in sea level were the result of a change in the rate of rotation of the earth caused by a brush with a celestial interloper. The renowned scientist Harry H. Hess suggested this in *Science* in 1936. Faster rotation of the earth would draw the earth's fluid inventory toward the equator. Often called the *Hypothesis of Ellipticity*, the higher rates of rotation would expose extreme northern and southern lands but submerge those near the equator. This hypothesis seems rather farfetched until one recalls that two miles of seawater represent only 0.05 percent of the earth's radius. Submarine canyons, however, are a worldwide phenomenon, occurring at all latitudes; the Great Bahama Canyon (25° N latitude) being one of the deepest. The canyons are too uniformly distributed to be attributed to a change in rotation.

Even more intriguing to the Catastrophist is the suggestion (by Elwood C. Zimmerman in *Science* in 1951) that great changes in sea level may be the result of vast outpourings of volcanoes and igneous rock around the world. For the uninhibited, this leads us back to the sea-floor spreading hypothesis, with great areas of water-bearing volcanic rock being exposed as the continents drift apart. The Catastrophic scenario might go something like this: Some great astronomical cataclysm involves the earth toward the end of the Cretaceous period; worldwide sea level drops precipitously as new ocean basins are opened up; as continental drift continues, huge quantities of water are released from the hot magma welling up along the midoceanic ridges. In this vision the submarine canyons would have been formed by water run off from the continents as the seas gradually filled up when water vapor was added to the atmosphere from magmatic outpourings from the earth's interior. Great changes in the earth's inventory of water seem more plausible than the simultaneous sinking of all continents as well as the midocean island groups; viz., the Hawaiian Islands where submarine canyons also exist. The concept of a drastically lowered sea level also squares with the belief that the observed distribution of fauna and flora could be explained by the exposure of land bridges and chains of seamounts, particularly the guyots that seem to have been wave-beveled during the Cretaceous period and later. As we have seen in preceding chapters, guyots are probably created by natural subsidence of the sea floors as they move away from the rift-associated spreading centers. In addition, Egon Orowan's recently proposed mechanism for sea-floor spreading depends upon the *absorption* of seawater by the upwelling magma along the rifts. Manifestly, submarine geology is far from being cut and dried.

Even the strongest adherents of Uniformitarianism have mused about the geologic revolutions that took place in the deep past when the seas came and went and when the very fabric of the planet was torn by internal and perhaps even extraterrestrial forces. Mountain building, convulsions of orogeny, sea encroachments; they are all part and parcel of conventional geology. The questions are *when* and *how fast* rather than *whether*. The canyon-cutting period is too recent for comfort. The last few hundred million years may have been placid and uneventful, but continental drift, magnetic reversals, and the submarine canyons tend to upset that notion. We live on a planet seething with contained energy; a planet that cruises through a solar system swarming with pieces of celestial flotsam and jetsam big enough to rend our globe to the core.

READING LIST

FAIRBRIDGE, R. W., ed.: *The Encyclopedia of Oceanography*, Reinhold Publishing Corp., New York, 1966.

HEEZEN, B. C.: "The Origin of Submarine Canyons," *Scientific American*, Aug. 1956.

JOHNSON, D. W.: *The Origin of Submarine Canyons*, Columbia University Press, New York, 1939.

SHEPARD, F. P.: "The Enigma of the Submarine Canyons," *Scientific American*, Sept. 1938.

―――― and Dill, R. F.: *Submarine Canyons and Other Sea Valleys*, Rand McNally & Co., Chicago, 1966.

7

BEYOND THE
GATES OF HERCULES

Ys, Lyonesse, St. Brendan's Isle, the Fortunate Isles, Atlantis; all
are dim remembrances of phantom lands handed down over the
centuries by legend and story. Many other legendary islands break
the long swells of the Atlantic and have appeared ghostlike
through the fog to Arctic adventurers. And, of course, there are
real Atlantic islands—the Canaries, the Azores, and the Madeiras
—and farther west the Americas are substantial enough.

The more we learn of the seafaring capabilities of the ancient
Greeks and Phoenicians, the more likely it seems that some of their
tales of Atlantic real estate may have some substance. The Greek's
Fortunate Isles, for example, were probably the Canaries. It is certain that the ancient Greeks knew of Britain and they probably
knew of Greenland, too. The ancient Egyptians may have actually
reached the Americas long before Leif Ericson, as Thor Heyerdahl
tried to prove with his 1969 transatlantic voyage in a papyrus boat.
How easy it would be to write this chapter if we were to point to
some piece of solid land—a volcanic spire on the mid-Atlantic
Ridge or vast shallows submerged within the past ten millennia—
and say this is a solid basis for the legend of lost Atlantis.

Atlantis is a shadow beneath the rolling sea; it is like a romantic
tale brought by whalers returning from the South Seas. In reality
Atlantis is merely a hint thrown up on our intellectual shores by

85

Plato more than two thousand years ago. Why Plato's few pages telling of a sunken continent and lost civilization should let loose a flood of thousands of books and ostensibly learned articles—mostly pro, a few con—is perhaps the real mystery of Atlantis.

Why Atlantis? Medieval fables tell us that Ys and Lyonesse were submerged in a twinkling like Atlantis when the sea rushed over portions of the Breton and Cornwall coasts. There is more "hard" evidence for submerged lands around the North Sea, where thousands of square miles lie just below the surface and human artifacts are continually dredged up from the bottom. For some reason, the more romantic of us want to believe in a less prosaic Atlantis; a great civilization that arose ten thousand years before Christ, prospered, colonized Europe and the Americas, waxed decadent, and was then destroyed by wrathful heavenly powers. It is like the Flood legend. Atlantis is a case history of man's futile attempt to reach beyond his God-given capabilities; Atlantis means far more to us psychologically than any Ys and Lyonesse awash in the North Sea.

Looking at the symbolic innuendos of Atlantis, it is not surprising to find who occupy opposing camps on the matter of Atlantis. The mystics, the romantics, and the occultists believe Atlantis was real to varying degrees. The Fundamentalists find little wrong with the basic idea, except that Atlantis seems to have held sway before Biblical creation of the world. Definitely in the opposition are most scientists. The scientists cannot find significant evidence—the whole tale is too romantic, too supernatural, and contrary to the scientific frame of mind. The question of Atlantis, then, is in a somewhat similar position to that held by the continental-drift hypothesis a decade ago, except that continental drift was never embraced by the occultists and pseudoscientists like the Atlantis hypothesis.

Now, how did the controversy get started? Plato was the main instigator, whether as a historian or fiction writer we will never know. About the year 355 B.C. Plato began a trilogy of three Socratic dialogues. The first was *Timaeus* or *Concerning Nature;* the second, which was never finished, was *Critias* or *Concerning Atlantis;* the third was never begun. Plato evidently intended *Timaeus* to be a sequel to *The Republic,* for it has the same cast of characters and dwells on the creation of the world and the nature of man. *Critias,* though, began a detailed description of the war between Atlantis and Athens, and the third book would probably have continued this story. According to *Timaeus*, the description of Atlantis was given to Solon, a statesman of ancient Athens, by an Egyptian priest when Solon toured Egypt six centuries before Christ. Solon himself is almost a legendary figure in Greek lore, and we have no way of knowing how much of *Timaeus* and *Critias* is due to Plato's imagination and how much is Solon's handed-down history with attendant distortion and embellishment. In other words, the origins of Atlantis are as foggy as the Atlantic itself sometimes is. But this is just the kind of grist from which Ignatius Donnelly constructed *Ragnarok* and Immanuel Velikovsky, *Worlds in Collision*. This, then, is not a very reassuring beginning.

In *Timaeus*, one of the four characters, Critias, sets forth the Atlantis proposition in a few straightforward words: Speaking of the exploits of Athens, he says:

> Many great and wonderful deeds are recorded of your state in our histories. But one of them exceeds all the rest in greatness and valor. For these histories tell of a mighty power which was aggressing wantonly against the whole of Europe and Asia, and to which your city put an end. This power came forth out of the Atlantic Ocean, for in those days the Atlantic was navigable; and there was an island situated in front of the straits which you call the Pillars of Heracles; the island was larger than Libya and Asia put together, and was the way to other islands, and from the islands you might pass to the whole of the opposite continent which surrounded the true ocean; for this sea which is within the Straits of Heracles is only a harbor, having a narrow entrance, but that other is a real sea, and the surrounding land may be most truly called a continent.

In the next paragraph Critias relates how the Athenian army triumphed over the invading Atlanteans. Soon after their victory their army was swallowed up in a single day and night of earthquakes and floods and "the island of Atlantis in like manner disappeared, and was sunk beneath the sea." In *Critias*, the second volume of the trilogy, some facts and figures about Atlantis emerge: it perished 9,000 years before Critias, putting the Atlantean empire back some 11,500 years from the present; the city itself was built on a circular plan and encompassed many ambitious works of engineering, such as a canal some seven miles long leading from the Atlantic to a protected harbor and an irrigated plain 10,000 square miles in extent. The subsequent details about the government of

Atlantis and the sins of its people have little bearing on the geologic truth of Atlantis. Plato's Atlantis was a well-embellished account of a real or fictional city; Plato gave Atlantis substance. But
in all that Plato wrote about Atlantis there is no clue that will help
us separate his fact from his fiction. It is a very cold trail upon
which we embark.

Atlantis has been promoted and disposed of by hundreds, more
likely thousands, of people from all scientific and humanitarian disciplines. It is also a valued property of the cultists, who venerate it
alongside other lost continents, such as Mu, the Atlantis of the Pacific. L. Sprague de Camp has charted the ups and downs of Atlantis stock in his book *Lost Continents*. The following highlights
are set down primarily to complete the Atlantis story rather than
for the small impact they have on the scientific reality of Atlantis.

Plato's pupil Aristotle implied that Atlantis was wholly fiction
(presumably he had some inside information), but Plato was highly
respected among the Romans and they took Atlantis more seriously. Under Pax Romana many cults flourished. The idea of a
great empire in the dim past that also ruled most of the world was
attractive. People *thought* about Atlantis but no one tried to verify
anything when the trail was somewhat warmer than it is today.
Christianity finally put an end to all the surmise about Atlantis
when it redirected the thoughts of men to the next world rather
than civilizations past. During the Middle Ages the idea of Atlantis
sank like the continent itself.

When America was finally discovered once and for all in 1492,
the vast western continents were quickly identified with the Atlantis of Plato. Sir Francis Bacon made America the scene for his uto-
pian *The New Atlantis*, thinking America a part of the Atlantean
empire that had not sunk. That the Americas were not the original
Atlantis was quickly decided when the aborigines turned out to be
too primitive, at least in the eyes of the European explorers, ever
to have built a great city and swept across Europe with their armies.

However, the fact that there were natives in America before the
white men arrived needed some sort of explanation. Were they the
result of a separate act of creation? Perhaps they were the descendants of ancient Phoenicians blown inadvertently across the Atlantic, or as some would have it, the progeny of Prince Madoc and his
band of Welshmen who were supposed to have crossed the Atlan-

tic in 1170. Not all of the American aborigines were primitive; the Mayans had books and knew astronomy. However, the Spanish conquerors were determined to eliminate all evidence of Satan's work and burned most of the Mayan books and many Mayans as well. Anything the Mayans knew of Atlantis went up in smoke. Nevertheless, the very sophistication and size of Mayan engineering works, particularly their cities, gave rise to the supposition that some Atlanteans had escaped earthquake and flood and made it to the Mexican shores of Yucatan. In this idea is the germ of the cultists' Atlantis—the Atlantis that was the "mother of empires."

To Plato the Atlanteans were merely an invading sea people with a taste for more land. To the modern Atlantist, who is aided by occult insights into the past, the inhabitants of Atlantis had nuclear power, miraculous metals, and great powers of war. All our civilization is ultimately derived from Atlantis. Perhaps, as Atlantists are wont to speculate, our technology will also run amok and our cities will slide beneath the waves just as mighty Atlantis did nearly twelve thousand years ago.

This vision of Atlantis, like the stories of little men in UFO's, has greatly impeded the search for any truth that might exist in the Atlantis legend. What reputable scientist will risk his standing by associating himself with obvious crackpots? None; there is no science of lost continents. Drifting continents, yes; sunken continents, particularly Atlantis, no.

In the preceding telescoped history of the Atlantis idea, Ignatius Donnelly was intentionally bypassed. This roly-poly mental giant, with his sweeping view of the cosmos past, was the first to gather up and consolidate all evidence hard and soft that might pertain to the reality of Atlantis into a single volume: *Atlantis, the Antediluvian World.* Before we look at the evidence of modern science, we must examine what Donnelly and his host of followers found in the way of indirect evidence, innuendo, and circumstantial correlation. *Atlantis, the Antediluvian World* is a classic example of the so-called "vacuum cleaner" approach frequently employed by pseudoscientists. It is important to understand the advantages and pitfalls of such an approach. Therefore, we will examine *Atlantis,* an archetype of the species, in some detail.

Proving the existence of Atlantis is not like proving the existence of an electron. With proper instruments any scientist who so wishes can duplicate the instrument readings by which physicists define the electron in terms of mass, electric charge, and so on. But there is no Atlantis at our fingertips, just as there are no sea serpents or UFO's at the disposal of science. In the case of sea serpents and UFO's we have large masses of reports of sightings with various degrees of validity; with continental drift and Atlantis the evidence is all indirect; that is, we see secondary and tertiary effects or worse. "Proof" of all four phenomena mentioned above customarily takes the form of a detailed cataloging of all pertinent information—and some not so pertinent. In fact, a comparison of Wegener's *The Origins of Continents and Oceans* with Donnelly's *Atlantis, the Antediluvian World* reveals many similarities in approach as well as many similar elements of "proof." Later in the book, Heuvelman's *In the Wake of the Sea Serpents* also will be found to depend upon a similar great mass of data, spanning centuries and derived from sources of varied degrees of repute. However, the difference between Donnelly and Heuvelman is in the discrimination exercised by the cataloger—and this makes the difference between science and pseudoscience. So much for the technique usually employed in proving elusive "things."

The first three chapters of Donnelly's *Atlantis* deal with Plato's story and the likelihood that it might be a valid account of a sea empire to the west of Greece. The next three chapters are the ones that Wegener seems to echo several decades later in his attempt to prove continental drift: "Was Such a Catastrophe Possible?" "The Testimony of the Sea," and "The Testimony of the Fauna and Flora." In support of the contention that continents can founder, Donnelly states: "There can be no question that the Australian Archipelago is simply the mountaintops of a drowned continent, which once reached from India to South America." There are numerous references to sudden subsidence of land in Iceland and elsewhere. Another Donnelly quotation: "According to Humboldt, along a great line, a mighty fracture in the surface of the globe, stretching north and south through the Atlantic, we find a continuous series of active or extinct volcanoes." It all sounds very modern and much like earlier chapters in *this* book, yet it was written in 1882. There was no question in Donnelly's mind that continents could sink beneath the sea and that the English *Challenger* Expedition had found remnants of Atlantis when it mapped portions of the mid-Atlantic Ridge.

"Proofs are abundant that there must have been at one time uninterrupted land communication between Europe and America." Donnelly then goes on to record many of the same similarities that Wegener and others noticed between the Americas and Europe and Africa. The difference between Wegener and Donnelly is that the former was trying to prove the validity of the continental-drift theory while Donnelly was accounting for the same similarities with a transatlantic land connection that suddenly foundered. Further, Wegener's connection between continents was supposedly broken 150 million years ago instead of the 11,500 years stipulated by Plato.

The rest of Donnelly's *Atlantis*—about five-sixths of it—deals with ancient legends from both sides of the Atlantic that seem to support the existence of a focal point of civilization, somewhere between Europe and the Americas, that one day sank beneath the waves. There are two "facts" Donnelly tries to prove in this, the major part of his book: (1) said terrestrial convulsion really did occur; and (2) culture radiated outward from what is now the Atlantic Ocean.

As Donnelly shifts gears from more or less conventional scientific topics to legends of floods and geologic turmoil, he sounds more like Velikovsky than Wegener. Donnelly did not have to look hard for flood legends; the inhabitants of lands around the Atlantic's shores preserve such records. The Biblical Deluge and the similar tale from Babylonia and Greece (Duekalion's flood) are best known. However, India, Scandinavia, and Indians of Central and North America have their versions of troubled times when the waters covered large tracts of land. Usually, the flood of legend was sent to punish mankind for wrongdoing. The only portions of the world that seem to be without flood legends are Australasia and interior Africa.

Do the Flood legends prove the reality of the Flood any more than *Worlds in Collision* proves the occurrence of an earthly flirtation with Venus? The basic nature of the data is identical, and both sets of data are persistent in folklore. "Proof" really comes only when a consensus of experts supports a theory. Manifestly, no expert can go out and duplicate the Flood as a scientist can with a laboratory-sized phenomenon. Proof therefore becomes a matter of persuasion. The data, such as they are, supporting the so-called Flood are so convincing that most scholars in the humanities and scientists agree that some inundations did take place in ancient times—at least on the scale of the Tigris-Euphrates Valley. The cause of the Biblical Flood and its extent are still subjects of contention.

Another sort of legend is even more intriguing than the circum-Atlantic Flood. Many ancient peoples around the rim of the Atlantic basin claim that their ancestors or "sophisticated" strangers came from the Atlantic. Ships and white visitors sometimes are mentioned specifically. Donnelly mentions the Basques, the Celts, the Mayans, the American Indians—all ancient peoples whose origins seemed curious and even remotely mysterious to him. These legends helped along the cause not only of the Atlantists but also of the so-called Keltomaniacs who believe that the Druids were the real creators of civilization rather than the Atlanteans. The promotors of the Lost-Tribe-of-Israel-in-America also draw on these ancient tales. L. Sprague de Camp filters out the misinterpretations and distortions from these legends in *Lost Continents*. Take the Aztec myth of Quetzalcoatl, for example. Quetzalcoatl was supposed to have been a ruddy-complexioned, bearded foreigner who arrived from the east. According to the story he went about Mexico with a group of black-robed followers spreading civilization throughout the land. He left on a boat of snakes promising to return. De Camp blames the early Spanish conquerors for distorting the myth by applying European makeup to Quetzalcoatl in their desire to make their conquest seem legend-come-to-life. In truth, Quetzalcoatl is generally shown as black-haired and black-faced in Aztec drawings.

The water gets even muddier when Donnelly attempts to prove cultural diffusion from Atlantis to both sides of what now constitutes the Atlantic Ocean. To do this, one must show similarities between primitive cultures in the Americas and Europe and Africa. Some things that men do, such as making fire or using a club, seem instinctive. We cannot prove the existence of Atlantis with anything like such common, almost obvious, inventions. If all humanity came from the same mold, even different races should have much in common. The believability of cultural diffusion from Atlantis really rests upon uncommon similarities, those correlations that are not so obvious. Some unusual practices that are often employed as proofs by the Atlantists are listed below:

(1) Circumcision
(2) Mummification

(3) Head deformation

(4) The use of the swastika

(5) The couvade, when the husband goes to bed and pretends to suffer the pangs of childbirth while the wife is actually doing the real job.

Some of the usual correlations proffered to prove Atlantis have been dropped from this list. For example, sun worship, the building of pyramids, brewing, the practice of witchcraft, the prevalence of the serpent in motifs, etc.; all of which seem pretty likely and understandable in almost any culture. But the couvade; there is a strange practice. Head deformation, too. In all honesty, though, it is all in the eye of the beholder. The essence of proof is in converting other people to your way of thinking regardless of their predilections.

In *Atlantis*, after relating how the couvade is found among North and Central American peoples, the Basques, the Corsicans, the Canary Islanders, and others around the Atlantic basin, Donnelly summarizes thus: "A practice so absurd could scarcely have originated separately in the two continents; its existence is a very strong proof of unity of origin of the races on the opposite sides of the Atlantic . . . ." Likewise, Donnelly shows the cultural ties of mummification, head deformation, circumcision, etc., to be "proof" of diffusion of ideas from some central focus. There seem to be many spokes to the cultural wheel, all ending on the shores of the Atlantic, but the hub seems to have disappeared. The hub must have been Atlantis. So goes the logic of the Atlantists.

Doubters of Atlantis have criticized everything in Donnelly's approach: the accuracy of his facts, the validity of the circum-Atlantic comparisons, his insistence that these cultural features diffused from a single source rather than being invented independently, and, of course, the believability of the whole logical structure. Unquestionably, Donnelly and most who followed in his footsteps leapt to conclusions and bent the facts this way and that to fit the hypothesis they were trying to prove. The cultural diffusion link in the Atlantis chain of reasoning is probably the weakest, but one is left with the nagging feeling that there are just too many strange coincidences and transatlantic connections to be explained away. An admission like this shows the power of the "vacuum cleaner" method of proof that Donnelly, Velikovsky, and others have employed so effectively; the mind tires of deflecting each missile (by explaining it in another way) and soon is ready to admit that the hypothesis might have some truth in it.

On and on Donnelly and his successors go, brick after brick in their Temple of Truth. The Mayan language is largely pure Greek says one Atlantist; it is similar to the language of the Basques says another. So it goes from cultural diffusion to language similarities to anthropology to motifs in architecture. The reader faces the mountain of "evidence." Where there is that much smoke must there not be some fire? This is a dangerous stance to take because the whole mountain of correlations cannot match the scientific value of one hard fact, such as, say, the dredging up of human artifacts from the mid-Atlantic Ridge. Quoting from De Camp's *Lost Continents:* "Most of Donnelly's statements of fact, to tell the truth, either were wrong when he made them, or have been disproved by subsequent discoveries."

De Camp does not and cannot dispose of all of Donnelly's claims; a great deal of mystery still exists about the origin and diffusion of cultures and peoples. Delicate surgery must cut away all suspect evidence, leaving only the facts that everyone can agree upon. This is what De Camp did in *Lost Continents* in 1954. He concludes that the Atlantis of Plato and especially the highly embellished Atlantis of Donnelly and the later Atlantists never existed. He believes that Atlantis was the product of wishful thinking: "So let us, once and for all, grant decent burial to all the theories of Atlantis that assume that a great island did, as Plato says, sink beneath the sea in a day and a night of storm and earthquake."

No one has the last word in science, particularly on its speculative frontiers. One needs only to read the advertisements in *Fate* magazine to find that the Atlantists are still alive and kicking vigorously. The books of "Colonel" James Churchward, such as *The Lost Continent of Mu,* have appeared in pocketbook form to expose everyone to a mutation of Atlantism, Muism. Mu is the Pacific Atlantis and the progenitor of the comic strip *Alley Oop,* a resident of "Moo." Further, the noted prophet Edgar Cayce in the pocketbook *Edgar Cayce—On Atlantis* predicts that Atlantis will soon emerge from the Caribbean Sea to become dry land once more. Then, perhaps, we'll be able to check out Plato's tale to the last detail.

While we are waiting for the resuscitation of Atlantis the myth of the drowned city is still with us. It is nagging and persistent; it refuses to die. Perhaps it is the increasing recognition by science that our planet's past has been a troubled one that makes the Atlantis catastrophe seem more conceivable. When De Camp wrote *Lost Continents*, the prevailing scientific opinion was such that he felt obliged to write off continental drift as a reasonable geophysical process. The scientific climate has changed, though, and it is now quite credible to slide continents around the globe to see how they fit. If one of the puzzle pieces were to settle a bit in the process, we could "lose" some dry land to the sea; although probably not as suddenly or as recently as Plato tells us. There are other valid mechanisms for disposing of terra firma. A large meteorite could have blasted Atlantis out of existence, as some Atlantists have proposed. This idea dovetails nicely with the legends of celestial turmoil that Velikovsky appropriated for his astronomical thesis. If Hudson Bay is a drowned meteor crater, a city the size of Plato's Atlantis would have been an easier excavation job. Finally, an island along the mid-Atlantic Ridge or in some other volcanically active region could have blown up leaving behind only memories of a sea-girt city that the gods destroyed one day. Let this suggestion that science now permits a slight resurgence of Catastrophism introduce the modern search for Atlantis.

Assuming we had the inclination, just where would a modern scientist look for more convincing evidence of Atlantis? The word "where" is used in its geographic sense here because further analysis of the ancient literature seems fruitless. What is needed is solid archaeological evidence for Atlantis, not more interpretations of tales that have been handed down and recopied through twenty or thirty centuries.

One could search along the mid-Atlantic Ridge with research vessels hoping to dredge up artifacts from the ruins of sunken Atlantis. What negative sport this would be, considering the thousands of miles of blind groping required. Happily, an alternative exists. The ancient Greeks were great exaggerators and the Mediterranean seemed pretty big to their small boats. Although Phoenician and Greek craft occasionally ventured well beyond Gibraltar, their ships generally hugged the coasts. Might it not be that Atlantis was really some city on the Atlantic coast, to the north in what is now Spain or somewhere along the ocean edge of Africa?

One surmise is that Atlantis and the old city of Tartessos (the

Biblical Tarshish) were the same. The Phoenicians contacted Tartessos as early as 1000 B.C. and found a rich city on a plain bordered by mountains. Tartessos was located near what is modern Cadiz, on the mouth of the Guadalquivir River. The area is not highly populated today, being a marshy expanse built of estuary silt. The location of Tartessos was not too far from where Plato placed Atlantis; it was rich and, like Atlantis, occupied a plain by the sea. Increasing the similarity is the fact that Tartessos seems to have suffered some natural calamity long before Plato's time, for it disappears rather mysteriously from the ancient records. Perhaps, as the Atlantis-in-Tartessos school believes, the river estuary subsided, wiping out the city and creating the shoals Plato said existed where Atlantis had sunk. Tartessos is not a phantom city. Archaeologists have explored it to some degree and confirmed its antiquity, but the high water table makes exploration difficult. Although a rich and important city once, Tartessos has little of the romance Plato built around his Atlantis. Needless to say, the archaeologists have found neither buried nuclear power plants or any of the precocious technology modern Atlantists claim for their city.

Some scientists are romantic enough to believe that Atlantis probably did exist—that there is some core of truth in Plato's story —but they compress the Atlantic Ocean into the Mediterranean and search for lost cities within narrower bounds. Ancient Crete and Carthage both have some similarities to legendary Atlantis. Carthage in particular was a great sea power with a well-planned city built on a circular plan. It was in the right direction from Greece too. Carthage never sank, though, and the geographical similarities are not too convincing. (The Atlantists, of course, claim that any resemblances between Atlantis, Crete, and Carthage exist because of cultural diffusion outward from Atlantis into the Mediterranean area.)

Ignore the Atlantists for awhile and continue exploring the Mediterranean for lost Atlantis, not as a lost continent but merely a lost city or island. In keeping with this de-exaggeration of *Timeas* and *Critias,* let us assume further that Plato for some reason put Atlantis too far back in time as well as too far away in distance; say, by a factor of ten for both. Atlantis would then be well within the Mediterranean and would have disappeared a couple of thousand years ago rather than a dozen millennia back. This assumption allows us to associate Plato's story with an actual natural catastrophe that apparently devastated the Mediterranean world about 1470 B.C. This was the explosion of the volcano Stronghyli, sixty-two miles north of Crete. The repercussions from this cataclysm may explain not only Atlantis but the Deluge of the Bible, the demise of the Minoan civilization, and the great climatic change around the shores of that inland sea.

For more than thirty years, Professor Spyridon Marinatos, inspector general of Greek antiquities, has been advancing the theory that the advanced Minoan civilization fell when a giant tidal wave engulfed Crete following the explosion of Stronghyli. The suffocating volcanic ash that rained down on Crete may have made agriculture impossible for years and further contributed to the Minoan demise. The explosion of Stronghyli was linked to the Atlantis legend by Professor Angelhos Galanopoulos, director of the seismic laboratory of the University of Athens. Galanopoulos first expounded his hypothesis at the 1960 meeting of the International Union of Geodesy and Geophysics. Here, at last, is an Atlantis hypothesis that can be checked out readily with all of the apparatus of modern science.

Oceanographic research ships from the United States and Sweden have discovered at least ten important layers of volcanic ash deposited on the floor of the Mediterranean. The ages of these layers vary from 3,500 to 80,000 years. Evidently the sea-bottom record will amply support all those legends of a turbulent ancient world, including Atlantis. Zeroing in on the Galanopoulos version of Atlantis, two ash layers seem associated with Stronghyli, the oldest and the youngest. The youngest, occurring about 1500 B.C., is the one associated with the Atlantis legend and the Minoan catastrophe.

Stronghyli, now called Thera or Santorin, provides something solid that scientists can work with. Even though the postulated eruption occurred 3,500 years ago, geologic evidence of the catastrophe still remains and is much more substantial than the words of Plato's works. Thera is associated with several smaller islands, which with Thera ring a sea-filled cavity some four miles in diameter. Two small islands break the water's surface near the center of the crater; one is volcanic and erupted in 1956. Estimates indicate that the volume of lava, dust, gas, and other vapors emitted in 1470 B.C. was four times that ejected by the famous 1883 eruption

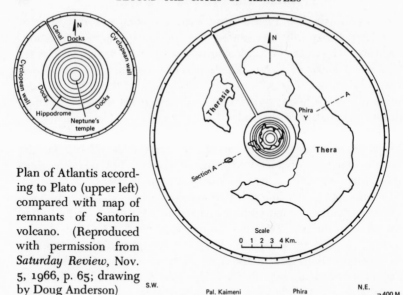

Plan of Atlantis according to Plato (upper left) compared with map of remnants of Santorin volcano. (Reproduced with permission from *Saturday Review*, Nov. 5, 1966, p. 65; drawing by Doug Anderson)

of Krakatoa. The cataclysm of Thera cannot be associated with any of today's muted mutterings of Mauna Loa; it was a true catastrophe of the first order.

Thera itself was buried under more than one hundred feet of ash. Ash and deadly vapors probably swept across the seventy miles to overcome Crete. There is physical evidence, too, of tidal waves more than one hundred feet high swamping Crete and smashing into Egypt and other territories bordering the Mediterranean. As a matter of fact, much of this physical evidence—silt and sand deposits—is the same as that used to support the reality of the Biblical Deluge.

Even more convincing is the archaeological evidence. In 1966, when soundings around Thera by James W. Mavor, Jr., on the research vessel *Chain* from Woods Hole Oceanographic Institution, showed geometrical similarities to Plato's descriptions of Atlantis, an archaeological expedition was organized. Led by archaeologist Spyridon Marinatos, the diggers found three-story houses, looms, vases, lamps, and the remnants of a large city under the ash. Capable of sustaining a population of perhaps 30,000, the city is roughly 3,500 years old and strongly resembles Minoan remnants found in Crete. Strangely, no human remains have yet been found, indicating that the inhabitants may have had enough warning to escape to other parts of the Mediterranean, carrying with them the relatively advanced Minoan culture.

The parallelism between Thera and Atlantis becomes more and more apparent; cataclysm, flood, people fleeing, cultural diffusion. Now it is a strange part of the whole Thera investigation that science, which often rejects evidence extracted from our literary legacy, takes the same evidence to support the Thera-is-Atlantis hypothesis. For example, old Egyptian documents plead, "O that the earth would cease its noise . . . . The towns are destroyed . . . . Upper Egypt has become wasted." The papyrus record tells of days of prolonged darkness when the sun shone like the moon. Scientists also say that a volcanic explosion like that of Thera could have caused the red rains, the red seas, the whirlwinds, and even the Ten Plagues mentioned in the Bible and other long-preserved documents. Professor Galanopoulos has even suggested that the tidal waves created by the eruption might first have drawn the sea back from Egyptian shores (often the first sign of a tidal wave is an initial *recession* of water) at just the moment when Moses and his people were fleeing before the pharaoh's armies. The Jews made it across the Red Sea bed, but the soldiers on their trail were caught by the onrushing tidal wave.

These ancient tales smack of Donnelly and Velikovsky. The difference is that Galanopoulos began with a hard fact, the eruption of Thera, and worked out the consequences based upon the known behavior of tidal waves and volcanic ejecta. Only in the end were the old tales of catastrophe brought in as supporting evidence. Donnelly and Velikovsky worked backward from the written record and ended up with the conclusion that astronomical disturbances had wrenched the earth. We have no good evidence of astronomical indiscretions within the solar system, but the eruption of Thera is well established.

To a scientist there is no question that Thera hypothesis is far preferable to the extreme claims of the Atlantists. Scientists and students of the ancient world are genuinely intrigued by the evidence around Thera. The eruption would explain in one stroke the sudden extinction of the Minoan civilization, the precipitous change in the Mediterranean climate, as well as many myths, fables, and legends that have perplexed them for millennia. The differences between the Thera and Plato's Atlantis theories are the location, size, and timing. The gulf between Thera and the Atlantist's Atlantis is infinitely wider. Perhaps Thera is only a propitious and convenient way to defuse the Atlantis issue. Now, whenever a scientist is asked about Atlantis, he can say with some assurance, "Yes, Plato's story is exaggerated but true because we have found the remains of Atlantis, just as the legendary city of Troy was finally discovered by the famed German archaeologist Heinrich Schliemann."

Have all the dreams of the Atlantists crumbled with the discovery of Thera's violent past? Not at all; the Atlantists will not deny that some local disturbance occurred north of Crete in 1400 B.C., but its scale was too small for their purposes. The circum-Atlantic correlations are not explained by Thera's explosion. A city of thirty thousand souls on a tiny Aegean island could not have been mighty Atlantis with armies (legend tells us) of millions. Clearly, we must look beyond Gibraltar for the true Atlantis, the Atlantis that was the "mother of empires."

Sailing past Gibraltar into the open Atlantic, there is little showing permanently above the surface except the Azores, the Madeiras, and the Canaries. All of these islands were apparently known to the Phoenicians. Geologists are still rather cautious about proclaiming their origins. The presence of small amounts of continental type rock among the predominantly volcanic foundations of the islands leaves the impression that perhaps they are debris left behind after continental drift or even continental foundering, somewhat like the continental Seychelles between Africa and India.

In 1958 the possibility of unknown continental remnants just west of Gibraltar were increased by the discovery of a submerged flat-topped bank off the northwest corner of Spain. Magnetic measurements and samples taken from four hundred fathoms indicate that this bank may be of continental origin. Underwater cameras, however, have shown no traces of past human activity.

It is not essential, of course, that Atlantis be associated with true continental formations. It could have been a large volcanic island. There are several such islands along the backbone of the mid-Atlantic Ridge—Tristan da Cunha, Saint Paul Rocks, etc.—but like Thera in the Mediterranean they seem too small for the grandiose version of Atlantis. In addition to the permanent, volcano-plagued Atlantic islands just mentioned, new ones seem to come and go. Atlantis might have been one of this ephemeral species.

In November 1963 the fishing vessel *Isleifur II,* working around the Vestmannaeyjar Island chain strung out southwest of Iceland along the mid-Atlantic Ridge detected heavy black smoke to the south. Upon investigation the crew of the *Isleifur II* were fortunate enough to witness the birth of a new island, which was soon named Surtsey. Island birth along the great crack down the center of the Atlantic is not uncommon, but how could a civilization like that ascribed to Atlantis ever take hold and prosper on the hot, unsteady top of a volcano? None of the up-and-down volcanic islands seem likely sites for an advanced civilization.

One of the strangest stories of the sea involves the fifteen-hundred-ton-steamship *Jesmond,* captained by David A. Robson (*American Mercury,* August 1956). In March of 1882 the *Jesmond* was bound for New Orleans from Messina with a cargo of dried fruit. Some two hundred miles south of the Azores the crew was startled to find the sea muddy and carpeted for miles with dead fish. Ahead, in water that was supposed to be two thousand fathoms deep, lay a smoking island. With a great deal of courage, Captain Robson led a landing party ashore. They were greeted by a vast barren plain of volcanic debris at the base of smoke-wreathed mountains. The plain was not as barren as it seemed, however. During the next two days the crew collected bronze swords, spearheads, clay vases carved with animals, and bone fragments. Masonry walls were found. Arriving at New Orleans, the collection was shown to a correspondent of the *Odebolt Reporter,* and a story was duly published on April 28, 1882. There was also a full account inscribed in the *Jesmond's* logbook, but this was destroyed in the Blitz against London during World War II. The collection of artifacts also seems to have disappeared, although Captain Robson stated that he was going to donate them to the British Museum. This disappearance is very suspicious in itself.

The story of the *Jesmond* and Captain Robson was quickly labeled a hoax by scientists, who happened to have been especially touchy since Donnelly's *Atlantis* had been published while the *Jesmond* was at sea. Hoaxes, though, are usually perpetrated by someone who has something to gain, even if only a good laugh. Captain Robson's story is said to have been verified by the entire crew, and the Captain himself went about his trade, never attempting to capitalize on his find. As for the island, it was never seen again; although in 1954 a new "bank" only eleven fathoms below the surface was reported near the position Captain Robson gave in 1882.

Now the *Jesmond* account has all the earmarks of a UFO sighting: something is seen by reputable observers but it is gone before experts arrive at the scene, if they choose to go look at all. The sea holds many mysteries of this undependable type. Among them are the two wartime sightings from planes flying between Natal, Brazil, and Dakar, in Africa. When over the mid-Atlantic Ridge, when the sun was at just the right angle, two pilots reported seeing the ruins of a town swimming tantalizingly just below the surface of the Atlantic. Captain Robson may have been an honest man, but the whole *Jesmond* tale has the earmarks of a hoax.

Two very recent undersea discoveries have excited the Atlantists and perplexed scientists. The first was the discovery of carved rock columns with writing on them in six thousand feet of water, fifty-five miles off Callao, Peru. The discovery was made accidentally by Dr. Robert J. Menzies, then at the Duke University Marine Laboratory, who reported his discovery in *American Oceanography*. While searching for mollusks with an underwater camera, Menzies photographed the columns, which were about two feet in diameter and sticking out of the mud five feet. A squarish block was also sighted. Although this discovery was in the Pacific and could hardly be Atlantis, the Atlantists will take lost continents wherever they find them. And, of course, the lost continent of Mu was supposedly in the Pacific.

The second find is even more exciting, although information is still lacking because the discoverers have refused to disclose the location of their find until they have exploited it. Sometime in 1968 Robert Brush, a cargo pilot, flying near the Bahamas, spotted the top of a "temple" just below the surface of the shallow waters of the Grand Bahama Bank. According to an Associated Press dispatch on August 23, 1968, the site has been explored by three Florida men: Dr. Manson Valentine, Dr. Richard Evans, and Dimitri Rebikoff. Valentine, a former zoology professor at Yale, stated that the top of the structure protrudes from the sand about two feet and that he has dug three feet farther down while inspecting the site.

Aerial views of the building have been displayed, and the discoverers are trying to obtain funds for an expedition to properly explore the site. Valentine reputedly pointed out that there may be some connection between the "temple" and the Atlantis legend or that it might be an outlier of the famous Mayan civilization.

It is difficult to judge this find in the Bahamas without more information. We do know, however, that there seems to have been substantial submergence of the land around the Bahamas. The Great Bahama Canyon, for example, can be interpreted as an indicator of great changes in sea level.

All of these hints are tantalizing; they keep the Atlantis story alive. Even a definitive study like De Camp's *Lost Continents* cannot inter Atlantis. Perhaps it should not; definitive studies have a way of being overturned by events. While the Atlantists are promoting their visions of a great Atlantean empire, a few scientists are beginning to apply the new tools of underwater research to archaeology. It is from our newfound ability to explore the sea bottom that any real proof or disproof of Plato's Atlantis will come. Atlantis will be with us for many more years.

READING LIST

DE CAMP, L. SPRAGUE, *Lost Continents*, The Gnome Press, Hicksville, N.Y., 1954 (A thorough, unprejudiced study).

DONNELLY, IGNATIUS, *Atlantis: the Antediluvian World*, Gramercy Publishing Co., New York, 1949 (Donnelly was the first Atlantist).

GALANOPOULOS, A. G., and BACON, EDWARD, *Atlantis*, Bobbs-Merrill Co., Indianapolis, 1969.

HEEZEN, BRUCE C., "A Time Clock for History," *Saturday Review*, Dec. 6, 1969.

LEAR, JOHN, "The Volcano that Shaped the Western World," *Saturday Review*, Nov. 5, 1966 (The Thera story).

MAVOR, JAMES W., JR., *Voyage to Atlantis*, G. P. Putnam's Sons, New York, 1968.

SPENCE, LEWIS, *The History of Atlantis*, University Books, Inc., New Hyde Park, N.Y., 1968 (Spence should probably be called an Atlantist).

8

THE WOMB OF LIFE?

After Darwin and his followers popularized the idea of life evolving from lower to higher forms in the last half of the nineteenth century, scientists turned their eyes toward the sea as the probable source of primeval life. Was not life incredibly abundant in the rich culture of seawater? Did not terrestrial life forms display similarities to more primitive sea creatures? Life undoubtedly came from the sea. It was so obvious that Walt Whitman wrote about his walks along the Long Island shoreline, "Where the fierce old mother endlessly cries for her castaways." In this line from *Leaves of Grass* Whitman expressed the consensus of science then as now: somehow, somewhere, life began in the great sea. This life advanced, eventually pulled itself out of the waters, became man, plus a million other species, and was now studying itself and its own beginnings and wondering what was to come next.

The origin of life is taught today as a perfectly natural, though more or less accidental, coming together of complex nonliving molecules in the sea, the "primeval soup." The standard version of the story goes on to imply that there is no great purpose behind these billions of years of evolution, that life is just another form of matter. In this view there is no destiny, no final goal. For those who cannot be happy with such an unromantic future, let us examine the "mystery" of life's origin in the sea—there may be additional hints of Catastrophism somewhere in the long unfolding of life,

104

Catastrophism that some may wish to interpret as acts of direct intervention in the process of evolution.

When one says that life "originated" at some time in the past, he infers that he knows how to distinguish between living and nonliving matter. One common definition of life states that any living form must: (1) metabolize; (2) reproduce; and (3) evolve or mutate. Most plants and animals obviously meet these requirements, whereas stones and television sets do not. The definition works in the main, but when viruses, crystals, DNA molecules, and other entities near the dividing line between life and not-life are examined, the choice is not so easy. There is no sharp dividing line. Crystals and DNA molecules, for example, grow and/or reproduce under the proper conditions. And it seems quite certain that DNA molecules, with their many millions of separate atoms, evolved from simpler molecules. Biochemists therefore speak of "chemical evolution" in the sense that simpler substances seem to combine themselves naturally into more complex structures, given the proper environment. Viruses apparently straddle the life/nonlife dividing line, leaning perhaps toward the life side. These pesky bits of matter may behave like nonliving crystals when separated and purified, but when the flu virus is let loose in the human body it reproduces too rapidly for the taste of most. Indeed, the "purpose" of a virus particle seems to be that of reproducing and expanding its domain, the same purpose as that of most animals it infects.

Even matter that is less lifelike than the viruses seems purposeful at times. Take liquid helium, He^4, which exists only near absolute zero. When the bottom of a test tube is pushed down into a pool of liquid helium, the fluid immediately runs up the side of the glass and fills the inside of the test tube to the level existing outside. When the tube is withdrawn, the fluid departs the tube via the same route. This is not siphon action. The climbing helium film may not be any more or less purposeful than a stone rolling down a hill under the action of gravity, but then human life may not be either. This, of course, raises the age-old question of determinism versus free will. Was it all preordained that the earth should coalesce; that the oceans should form; that life should begin and evolve; and that you should read this book?

More books have been written about determinism and free will than the existence or nonexistence of Atlantis, but this will not be one of them. The only points made in the preceding paragraphs are: (1) life and nonlife do not seem to be separated by a well-defined dividing line—there is no measurable "breath of life"; and (2) there are already intimations that once again we have a subject where one's view of the "truth" is colored by his philosophical leanings. For instance, the Russian scientist A. I. Oparin views life merely "as a special form of the motion of matter, which differs in qualitative terms from objects in the inorganic world." Life is merely a different kind of matter, a more evolved and therefore advanced form of matter. There is nothing fundamentally different along the long road of evolution from atom to molecule to life. But what lies beyond life as we see it today? Has evolution's end been reached at last?

Regardless of what the mystic or religious individual believes about any "higher purpose" of life, the fact is that science proceeds on the assumption that nonliving matter can, under the proper conditions, transform itself into living matter.

With more and more sophisticated machines and smarter and smarter computers being built every day, the debates about *reductionism* have become popular. A reductionist believes that living organisms, including man, can be described by the same laws the physicist and chemist use in describing simple, nonliving systems. No reductionist claims this can be done today, but he will argue forever that the principle is valid.

The possibility exists that life/matter has always existed throughout an infinite spectrum of simplicity-complexity, purposefulness-nonpurposefulness, or whatever criteria distinguish the different steps in evolution's ladder. Life is all around us in this view, which many will recognize as an expression of *panspermia,* that supposition that life on earth was seeded from outer space, perhaps by some wayward meteorite carrying spores. During the nineteenth century Lord Kelvin and Hermann von Helmholtz popularized this thought that a reservoir of life exists on bits of matter out in space. Even the great Swedish chemist Svante Arrhenius was attracted by panspermia, believing that light pressure carried spores from star system to star system. In recent years, however, the high radiation levels in space have damaged the panspermia hypothesis severely; it is now rejected by most scientists, even though some surprising lifelike forms have been found in the so-called carbonaceous chondrite meteorites (more on this toward the end of this chapter). Furthermore, radio telescopes have recently detected the spectra of ammonia and formaldehyde in outer space, indicating that even interstellar space boasts an impressive inventory of rather complex, life-oriented molecules.

If life did not come to the earth from elsewhere, it must have originated here. As mentioned before, science favors natural over supernatural creation overwhelmingly. The modern scientist would say that life *must* have originated here on earth. To those who argue that the chance coming together of just the right molecules in a primitive chemical soup is infinitely unlikely, the famous British biologist, Julian Huxley, once replied: "We can today assert, without fear of contradiction, that the argument from improbability has lost whatever validity it once possessed." The great age of the earth and the ease with which complex organic molecules were made in the laboratory prompted Huxley's optimism. Whatever Huxley may have said about it, we know only that life *may* have arisen spontaneously on earth at least *once*. Research on the origin of life today is aimed at:

1. Proving that life can be created from nonliving chemicals under the proper conditions. In essence this is an attempt to prove that the spontaneous creation of life is not improbable and that no vital force or "breath of life" is needed.
2. Demonstrating that the proper conditions are similar to those that probably existed in the earth's primeval seas.
3. Demonstrating that the fossil record shows continuous evolution from simple chemical substances to more complex chemicals and ultimately to the simple forms of life, to man and other creatures.

If life is created successfully by man in his laboratories, some will argue that it was possible only because the atoms and molecules, and the forces that hold them together in certain ways, were made that way on purpose by a supreme being. In effect, creation is moved from the "breath of life" moment back to the time when the intrinsic properties of atoms and molecules were established. The thought that life is a pattern of atoms and molecules like snowflakes and crystals is merely another way of stating Oparin's thesis that life is merely another statement of matter.

A strange philosophical cycle appears when we look back on the history of man's attempts to explain the life around him—life that is so fecund and ubiquitous. From the ancient Greeks until the time of Pasteur's experiments in the 1860's it was taken for granted that life was generated spontaneously from nonlife. Now, a century after Pasteur showed that such could not happen, mankind is again exercising all his ingenuity to prove that it not only can happen but that it is a perfectly natural phenomenon.

Democritus, the father of the atomic concept, believed that living creatures arose when the finest particles of moist earth encounter and combine with the atoms of fire. Aristotle went along with this basic idea; so did all the thinkers of the Middle Ages who hung on Aristotle's words. In the thirteenth century Saint Thomas Aquinas admitted the possibility of autogenesis of such animals as worms, frogs, and snakes in putrefied materials under the action of the sun; although man, of course, was created directly by God. Certain natural conditions always created life according to Descartes—a most modern thought. So it went: everybody talked about and believed in the spontaneous generation of life. After all, so many people had seen worms, maggots, and many lowly forms of life suddenly appear where they had not been the day before.

After many a "careful" experiment observing worms, maggots, frogs, and the other animals emerge from mud, slime, and putrescence, a few natural philosophers of the eighteenth century attempted something more deliberate and controlled. For example, the English priest and naturalist John Needham demonstrated autogenesis by taking hot mutton gravy and hermetically sealing it in a closed vessel. The vessel was then heated further in the hot ashes to kill any indigenous organisms. Several days later, Needham found that the "sterilized" gravy was teeming with microbes. Needham's conclusion: the spontaneous creation of life in decaying organic substances is not only possible but "obligatory."

Needham's mutton-gravy tests were typical of "life" experiments in the middle 1800's; spontaneous creation was the easiest thing in the world to reproduce. However, some scientists controlled their experiments much more carefully. The Italian investigator Francesco Redi in 1668 showed quite conclusively that rotting meat would not generate maggots if adequately protected from flies by cheesecloth. Further, if the fly eggs were allowed to fall on the meat, lo, the maggots again appeared. Redi's work convinced few; he himself continued to believe in other types of spontaneous generation.

Another Italian, Lazzaro Spallanzani, a contemporary of Needham, was of Redi's persuasion. Spallanzani sealed and roasted all manner of organic concoctions and showed that Needham had not been diligent enough in his sterilization procedures. Nevertheless, the world remained convinced of the truth of autogenesis.

Perhaps the last great confrontation between experimenters with opposing notions occurred in Europe while America was engaged in its Civil War. The scientific battleground was France. In 1859 Félix Archimède Pouchet published articles describing his experiments on autogenesis. He used infusions of hay that persisted in growing microorganisms even after the intense heating. Hay seemed a pretty lifeless commodity, and Pouchet's experiments persuaded the French Academy of Sciences to offer a prize for the one who could explain in the laboratory just how life coalesced and took form in experiments such as this. In 1862 Louis Pasteur described his extremely thorough work on autogenesis, which was patterned after Spallanzani, but with better safeguards. Coupled with the precision and conclusiveness of Pasteur's experiments was the discovery by John Tyndall, in England, that hay contained spores, a much hardier form of life that could not be killed by the techniques of Pouchet. Spontaneous generation was disproved almost overnight. After Pasteur, anyone who claimed to have observed spontaneous generation was thought (and often proved) to be a sloppy experimenter.

Oparin compares Pasteur's feat with that of Copernicus. Natural science was propelled into a crisis, for if life could not generate spontaneously, from where did it come? It must be divinely created; but divine intervention was quite contrary to the spirit of science. Science had one reasonable alternative: show that a different kind of spontaneous generation existed.

Before moving into more modern times, a fascinating sidelight of Pasteur's work (unpublished) was his attempt to isolate viable bacteria from a carbonaceous meteorite. He failed; but the fact that he tried shows the philosophical importance to many scientists of demonstrating that life did not descend willy-nilly upon the earth in some chance meteorite in panspermia fashion.

Indeed, the modern scientific tenet that simple molecules evolve into more complex molecules and thence into simple living forms seems to fly directly in the face of Pasteur's findings. In this modern view Needham's gravy experiments were poorly controlled; Pasteur's were technically commendable but not conducive to the slow, painstaking building of large molecular structures that must precede the creation of test-tube life. In a sense we are back to the gravy stage, but the raw materials come not from mutton but from the simple elements present in the primitive ocean and atmosphere.

A harbinger of modern sentiments arrived in 1909, when Professor Carl Snyder, in the United States, proposed that life originated in volcanic eruptions long ago when favorable combinations of chemicals and temperatures occurred accidentally. Today we harness ultraviolet light, lightning, and nuclear radiation to nudge molecules into more fruitful chemical associations. Ernst Heinrich Haeckel, the famous German naturalist, and others had formulated the "stimulus" concept during the late 1800's. It was generally held, however, that the conditions for such "magic moments," when everything was just right, no longer prevailed.

Again in 1909 Robert K. Duncan wrote an article for *Harper's Magazine* entitled "The Beginning of Things." Except for his imagery (most unusual for a scientist), Duncan might have written today.

> He [the man of science] cannot believe that there was actually a break between the inorganic and the organic evolutions bridged over by the direct action of the finger of God. He must believe that there has been no break whatever—the waving palm trees, the toddling children and wave-beaten rocks are alike, the present natural outcome of and absolute sequence of cause and effect, passing back to the blazing star that fanned the elements that comprise them. He must believe this because he believes the Law of Continuity—the Law of Laws.

This quotation is an appealing expression of Uniformitarianism, a statement of geologic and biologic evolution, a passionate rejection of Catastrophism. In a nutshell, this view holds that there is no intrinsic difference between living and nonliving matter.

The tack science currently plies was set to a great extent by Oparin, who published his first version of *The Origin of Life* in Moscow in 1924. Subsequent Russian editions in 1936, 1941, and 1957, as well as editions in other languages, attest to the influence of Oparin's thinking. In 1929 the English biologist J. B. S. Haldane presented his very similar ideas about chemical evolution and its step-by-step approach to life in his classic text also entitled *The Origin of Life*.

A fit conclusion to this historical introduction to "life" experiments is from Oparin:

To be sure, we are presently unable to detect the primitive origin of life in the natural conditions surrounding us, since the evolutionary process of carbon formations has an irreversible, unidirectional nature. We are also incapable of artificially reproducing this entire process as a whole in the form in which it arose in nature, since it spread over more than a billion years, but the separate stages in this process are quite accessible to objective scientific investigation.

Given this challenge, how have biologists and other life scientists answered?

The 1920's saw the beginning of what we might call Type 2 Life Experiments. The early Type 1 Life Experiments of Needham, Spallanzani, and Pasteur were aimed at proving or disproving the hypothesis of autogenesis in materials at hand, particularly organic, close-to-life substances. Now, the target was the deliberate step-by-step creation of life using biochemical knowledge. As Oparin pointed out above, we may never be able to duplicate nature's steps with precision all the way from carbon atom to amoeba, but we should be able to trace out most of the sequence.

The earliest experiments attempted to reproduce conditions that science believed existed soon after the earth was formed. At first the planet was pictured cooling after a molten birth; now, cold accretion is postulated, although the sun and radioactivity (a la Joly) may have created high temperatures during the past. In April 1923 Oscar Bandish reported to the American Chemical Society that he had produced typical organic compounds from air, water, and carbon dioxide simply by mixing them and exposing them to ultraviolet light. Similar experiments were performed with increasing frequency. Some, of course, lacked proper controls. Carelessness led to premature announcements through the years. Typical was the "discovery" of protoplasm and lifelike forms after inorganic matter had been exposed to light by the Mexican biochemist Alfonse L. Herrera in 1930. (The lifelike forms bore considerable resemblance to the "organized elements" found today in carbonaceous meteorites.)

In 1951 Melvin Calvin and his associates began what might be called the "modern" series of prebiological chemistry. These researchers merely exposed carbon dioxide and water to the particles in a cyclotron's beam, which simulated space radiation on the primitive earth. Significant yields of formaldehyde and formic acid were discovered after the bombardment.

In 1952 Harold C. Urey, an American Nobel Prize winner in chemistry, argued that the atmosphere of the young earth probably contained hydrogen, ammonia, water vapor, and methane. Urey thought that with the addition of energy, life might have got started with these basic ingredients. Following Urey's suggestion, one of his students at the University of Chicago, Stanley L. Miller, prepared a sterile brew of water and the suggested chemicals. An electric discharge in the sealed container provided the energy source. A few days later Miller detected a wide variety of organic molecules, including a few amino acids, in his primordial soup. This experiment received considerable publicity.

Since these pioneer experiments, dozens of experimenters all over the world have tried their own hand at building "microenvironments" simulating that of prebiological life. At first the primary ingredients were quite simple—water, carbon dioxide, ammonia, etc. It turned out to be almost too easy to make complex organic compounds. It was almost as if the simple molecules were "predestined" to become building blocks in more ambitious structures; all they needed was a little encouragement from an energy source. More complex reactants were tried such as malic acid, amino acids, urea, and formaldehyde. In other words, by using the products of the first simple syntheses, it was hoped that bigger molecules closer to life would be created. The energy sources varied widely: particulate radiation, ultraviolet light, heat, and even friction. The latter experiment was performed, using artificial projectiles, by John J. Gilvarry (see chapter 1), who proposed that the ocean basins were gouged out by meteorites—a process that should have entailed considerable friction. In 1965 Gary Steinman, working with Calvin at Lawrence Radiation Laboratory, found that prelife compounds even formed on wet earth at room temperature. The years have been the synthesis of ever more complex organic molecules as experiments climbed to the second step of the chemical evolution ladder. Monosaccharide sugars, the purines and pyrimidines of the nucleic acids, adenosine triphosphate (ATP), and the porphyrins, all of which are closely associated with life processes.

No question remains about the capability of science to push chemical evolution to the brink of life. On one hand we see that natural energy sources can push simple chemicals up the evolutionary ladder; on the other we find more deliberate biochemists, using all the "unnatural" methods of modern science, synthesizing DNA and large organic molecules several steps further up the ladder. Whether on some magic day some white-coated scientist will succeed in crossing the nebulous barrier between nonlife and life is still conjecture. The scientific consensus at the moment is that some day this last step *will* be made. There is even a faint suspicion growing that this steep ladder of evolution, the one we have made so much of, is in reality a natural course taken by all matter. The more speculative scientists talk of a trend toward life that is equal and opposite to the trend toward death imposed on the universe by the Second Law of Thermodynamics, which maintains that the universe is running down like a windup toy.

The "creation of life in a test tube" has become a popular subject in recent years. The reader will find several popular accounts of these experiments and their chemical details in the references listed at the end of this chapter. The key factors bearing on this and future chapters are: (1) the role of radiation, generally considered to be an important one; (2) the part played by the primitive oceans; and (3) the philosophical similarities between the principle of evolution (which includes crossing the nonlife/life barrier) and the geological principle of Uniformitarianism.

Before examining the geologic record of early life, it will be worthwhile to recapitulate how life originated on earth according to the current consensus.

The total scheme of evolution (see page 114) takes us from simple cosmic hydrogen down several billion years to man himself in an unbroken chain of increasing complexity. It is customary to show this "flow chart" vertically, like an industrial production line accepting raw, unsophisticated hydrogen at the top and discharging completed birds, quadrupeds, and primates at the bottom. Inside the factory there is a universe, in particular a planet called earth, which acts as a vast chemical plant. The primitive elements and compounds created during nuclear and molecular evolution are stirred by the planet's winds, cooked in its seas by a radioactive heat, and energized by solar and cosmic radiation. The major steps along the assembly line are:

1. The synthesis of very simple organogenic molecules such as methane, ammonia, carbon dioxide, etc., which are the precursors of organic compounds. Here the assembly line has not necessarily proceeded to

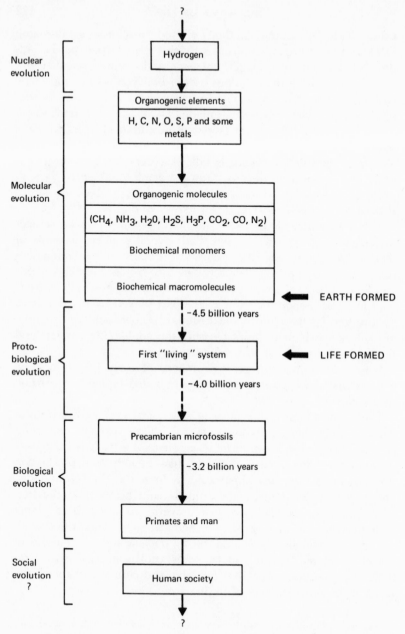

The "natural" evolution of inert matter into life. We do not know how long this chain is or even whether there is an end.

114

the point where the earth has been formed, rather planet-stuff can be manufactured in stellar atmospheres and outer space.

2. In the second stage the simple organogenic molecules evolve into true organic compounds.

3. In the last nonlife stage, the biochemical polymers build themselves up from the monomers. For this synthesis it is usually assumed that the earth has formed, with its liquid envelope and conducive temperatures, chemicals, and radiations. In other words a planet similar to the earth is essential to the evolution of life as we know it.

4. Within a half billion years after the earth's formation (4.6 billion years ago) the first continuous self-replicating molecular system appeared in what is usually termed protobiological evolution. Several substeps are recognized: (a) the appearance of the first "autocatalytic" molecules (those that stimulate their own multiplication); (b) the emergence of the first informational self-duplicating molecules; (c) the appearance of the first information-translating molecule; and (d) the formation of a system of molecules that form a stable boundary between phases. The four physical entities corresponding to the substages above are (a) enzymes; (b) DNA (deoxyribonucleic acid); (c) transfer-RNA (ribonucleic acid); and (d) the cell membrane. These "events" may have been the most significant chain of actions ever to occur in the universe—in essence they represent the "breath of life."

In a very perceptive article in the January 7, 1955, issue of *Science* (vol. 121, p. 9), the American biologist Hermann Joseph Muller concluded with the following paragraph, which nicely expresses a scientist's view of this tendency of matter to come together to make life:

Who can say how far this seed of self-awareness and self-transfiguration that is within us may in ages to come extend itself down the corridors of the cosmos, challenging in its progression those insensate forces and masses in relation to which it has seemed to be but a trivial infestation or rust? For the law of the gene is ever to increase and to evolve to such forms as will more effectively manipulate and control materials outside itself so as to safeguard and promote its own increase. And if the mindless gene has thereby generated mind and foresight and then advanced this product from the individual to the social mind, to what reaches may not we and our heirs, the incarnations of that social mind, be able, if we will, to carry consciously the conquests of life?

With this ringing declaration by Muller (a Nobel Prize winner in 1946) about the past and future "conquests of life," let us now look at the fossil record of a great conquest, perhaps the greatest conquest of life: the establishment of a firm beachhead on the planet earth, possibly the only spot in the universe where life has taken root.

Despite the supposed inevitability, early life forms must have had tough going on the primitive earth. Atlantis-style inundations, widespread meteorite bombardment, vulcanism, and violent storms must have been common. Yet life did take hold billions of years ago according to radioactive dating and fossils found in sedimentary rocks laid down on the margins of the ancient seas. Analysis of the structures of these fossils and their chemical constituents provide the only real record of what happened billions of years ago. Even so, the evidence proving the fragile link between nonliving and living matter must have been destroyed a long time ago. There is no known series of strata that preserves a record showing the actual chemical-biological steps that occurred during the creation of life. It is perhaps the most vital "missing link" in the totality of chemical-biological evolution. This link must be inferred from laboratory experiments, although eventually the drama may be found enacted in the geologic theater.

Of late there seems to be an undeclared race in which paleochemists, paleobiologists, and other paleo-disciplines compete to find evidence of life in the oldest possible rocks on earth. All of which, of course, were deposited in shallow seas or bodies of fresh water. Few geologic records of life exist showing what happened on dry land or in the deposits of whatever oceans existed then. Almost every few months an announcement from some institution tells how life is really a few hundred million years older than we had imagined last month. Gradually, the beginning of terrestrial life is being pushed far back into the earliest history of the earth. The oldest sedimentary rocks seem to show that life existed more than three billion years ago. Sedimentary rocks older than this are metamorphized to an extent where clear-cut chemical evidence has been destroyed long ago.

Two different techniques are used: (1) sedimentary rocks are examined visually for structures that look like they might be the remains of primitive creatures; and (2) sedimentary rocks are studied in the chemistry laboratory where life-associated chemicals are identified. In both cases, of course, the rocks are dated by radioactivity. (Geologists must avoid the circular logic of dating rocks by their contained fossils and then dating fossils by the rocks in which they are found.)

Structural analysis of life forms is classical paleontology; and it is very successful with more advanced organisms, but there are problems where Precambrian rocks are studied. The remains of the first living things, when altered by heat, pressure, and chemical action, may resemble natural chemical structures that were never associated with life at all; viz., the so-called carbon snowflakes. Then, too, no matter how much care is taken, there is the possibility of contamination by pollen and microscopic forms of life from the living sea of organisms in which we are immersed.

The study of "chemical fossils" is also clouded to some extent by the effects of billions of years of heat, pressure, and chemical action. Despite chemical fragility, despite aeons of heaving terrestrial crust, paleochemists have found amino acids, peptide chains, and even proteins in well-protected spots, such as between the thin sheets of crystal. Even when the structure of the original organism has been destroyed, its chemical remains may pinpoint its biological class, perhaps someday even its species.

One by one the earth's most ancient sedimentary rocks have been found to be fossiliferous. In most reported cases, specific structures are highlighted as the most conclusive evidence, although chemical studies are also positive, as positive as they can be. The following table summarizes some of these recent discoveries.

FOSSIL FINDS IN ANCIENT SEDIMENTARY ROCKS

SEDIMENTARY ROCKS	AGE	PARTICULARS
Gunflint Chert (Ontario)	1.9 billion years	In 1965 E. S. Barghoorn and S. A. Tyler report the discovery of microorganisms.
Soudan Shale (Minnesota)	2.7	In 1965 P. E. Cloud et al. report microstructures of possible biological origin in this carbon-rich formation. M. Calvin detects phytane, a derivative of chlorophyll in same rocks.

SEDIMENTARY ROCKS	AGE	PARTICULARS
Fig-Tree Series (South Africa)	3.1	In 1967 E. S. Barghoorn and J. W. Schopf find 22 amino acids. Infer that basis of life has not changed materially in 3.1 billion years. Minute bacterium-like structures also observed.
Onverwacht Series	3.5	In 1968 A. E. J. Engel et al. report finding alga-like forms in what are probably the oldest sedimentary rocks on earth. These carbon-rich rocks are interlayered with thick layers of lava.

As far back as we can go we find evidence of life, both structurally and chemically. The day is not distant when one can say that all sedimentary rocks known on earth show some evidence that life existed when they were deposited. Life indeed is an ancient, all-pervading force.

The evolutionist explains the age and pervasiveness of life in terms of the speed and facility of his well-oiled, but mindless machine of evolution, which processes all matter from interstellar hydrogen atom to terrestrial primate (there may be more to the spectrum, but that's all we know now). As persuasive as the evolutionist is, who can deny that things are the way they are because these things had survival value.

The most interesting rock samples are those that fall from the skies, more specifically the so-called carbonaceous chondrites, which comprise about 3 percent of all bona fide meteorites. The carbonaceous chondrites not only display a spectrum of complex hydrocarbons but also lifelike structures called "organized elements." Only a few dozen of these rather rare meteorites are known, and all were observed when they fell. Pasteur and chemists before and after remarked upon the high organic contents of these specimens from outer space. It was not until 1961, when B. Nagy, W. G. Meinschein, and D. J. Hennessy reported on their studies on carbonaceous chondrites, that the controversy began in earnest.

These investigators asserted that the chemistry and organized elements they found in these bits of stone might be remnants of life. The storm rose; life forms could not possibly exist in meteorites. It was all due to poor science and terrestrial contamination of meteorite samples that had been lying around museums for decades. Though the opposition grew heavy, more and more scientists began taking carbonaceous chondrites apart and studying their supposedly uncontaminated hearts, under the microscope and in the test tube.

Harold C. Urey, a Nobelist in chemistry, summed up his views in an article in the January 14, 1966, issue of *Science:* "If the materials discussed here were of terrestrial origin, it would be firmly suggested that the materials were of biological origin and indigenous to the samples." More recently, in 1968, Gerhard O. W. Kremp, at the University of Arizona, rejected the claim that all of the puzzling apparently biogenic structures and chemicals in the meteorites are due to terrestrial contamination and, more sensationally, that some of the organized elements may be fossils of blue-green algae. Kremp has even carried out experiments duplicating the conditions of fossilization with blue-green algae and found that their cysts assume shapes similar to those of the organized elements. Needless to say, the whole question of life forms in meteorites is far from settled; the organized elements may be true fossils, terrestrial contaminants, or structures not derived from life at all.

Urey's remark that carbonaceous chondrites would be considered to contain material of biological meteorites if they had not fallen from the sky stimulates several fascinating questions:

(1) If the carbonaceous chondrites are truly terrestrially contaminated, why cannot the same reasoning be applied to the old sedimentary rocks from South Africa, which contain the same sorts of chemicals and structures? Perhaps our whole effort in paleochemistry is distorted by the same sort of contamination that bedevils the meteorite analyzers.

(2) If the carbonaceous chondrites are truly terrestrially contaminated, why could not these meteorites be fragments of the earth itself or perhaps the moon? Or perhaps the meteorites, the moon, and the earth were consolidated into a single mass at one time, as proposed by Osmond Fisher, George Darwin, and so many others. If the latter is the case, the organized elements in meteorites *should* appear to be

terrestrial contaminants because *they are*. Another theory of lunar genesis—that of lunar capture—infers such terrestrial havoc that small pieces of the earth may have been wrenched into orbit by large meteorites after the fashion of the suborbital tektites. These chunks of debris would have reentered eventually, blasting out large terrestrial craters.

The two theories, panspermia via extraterrestrial meteorites and the terrestrial origin of carbonaceous chondrites contain nothing impossible—just improbable. The bulk of the scientists hold with the theory that life evolved naturally, and, in the best tradition of evolution and Uniformitarianism, this is by far the most likely hypothesis. If the Apollo astronauts had returned with lunar rocks containing organized elements with innuendos of terrestrial life forms or, even "worse," legitimate terrestrial fossils, we would have had to review our life-origin and earth-moon theories again. Of course, the astronauts would have to assure us that they did not pick up carbonaceous chondrites, which must also strike the moon regularly. Fossils are customarily found in sedimentary rocks on earth. Although there are some photographic hints of water erosion on the moon, the actual discovery of sedimentary rocks there would be rather startling—unless, of course, the moon borrowed some from the earth during the postulated planetary fission.

In this chapter we have found that the main theme played by paleobiologists and paleochemists has had the slow, steady, rather bland beat of Uniformitarianism. Philosophically speaking, evolution, whether chemical or biological, is the blood brother of Uniformitarianism. Catastrophism acting through such speculative and spectacular events as the birth of the moon, meteorite impacts, and radiation storms may have helped bring life to the earth and stimulate its evolution; or, conversely, spread terrestrial life outward. But like all Catastrophic theories, these are considered weakly substantiated by evidence. Besides, no proponent of Uniformitarianism will deny that *some* catastrophes occurred over the aeons; he merely questions their significance in the total picture of life's origin and evolution. Further exploration of the earth's seas and the surfaces of the moon and planets will help us determine where the truth lies between these rather artificial philosophical extremes.

READING LIST

ABELSON, P. H.: "Paleobiochemistry," *Scientific American*, July 1956.

BRIGGS, M. H.: "Evolutionary Biochemistry," *Journal of the British Interplanetary Society*, June 1968.

EGLINTON, G., and CALVIN, M.: "Chemical Fossils," *Scientific American*, Jan. 1967.

FOX, S. W., and MCCAULEY, R. J.: "Could Life Originate Now?" *Natural History*, Aug.–Sept. 1968.

—— "How Did Life Begin?" *Science and Technology*, Feb. 1968.

JACKSON, F., and MOORE, P.: *Life in the Universe*, W. W. Norton & Company, Inc., New York, 1962.

OPARIN, A. I.: *The Chemical Origin of Life*, Charles C. Thomas, Publishers, Springfield, Ill., 1964.

—— *The Origin and Initial Development of Life*, "Meditsina" Publishing House, Moscow, 1966. Available in English translation as *NASA TT F-488* from the Clearing House for Federal Scientific and Technical Information, Springfield, Va., $3.00.

PLATT, R.: "When Life on Earth Began," *Reader's Digest*, Dec. 1961.

SIMPSON, G. G.: *The Meaning of Evolution: A Study of the History of Life and Its Significance for Man*, Yale University Press, New Haven, 1967.

UREY, H. C.: "Biological Material in Meteorites: A Review," *Science*, Jan. 14, 1966.

WALD, G.: "The Origin of Life," *Scientific American*, Aug. 1954.

9

CRISES IN THE
EVOLUTION OF LIFE

For more than three billion years life on earth has progressed from
the simple to the complex; from fish to fowl; from single cell to
man. This progression has been, the paleontologist says, a random
and opportunistic sort of trip during which life reshuffles its deck
of genes in response to environmental pressures and moves in to
exploit new or modified ecological niches. The mosquito adjusts to
DDT and man extracts power from uranium. Just as a drop of
bright red dye diffuses throughout a beaker of clear water, so life
randomly and purposelessly fills as much of the universe as it can
with its host of fertile, adaptive species.

All this talk of evolutionary pressure is pure speculation, of
course. Although one is tempted to ask whether automobile acci-
dents, lung cancer, radioactive fallout, and other modern environmental pressures are in any way modifying that reservoir of genes
—that huge stack of biological IBM cards coded with all the characteristics of the human race. Modern medicine, for example,
changes men's gene inventory by permitting children who would
have died from birth defects a century ago to reach child-bearing
ages. We really cannot tell what is happening over a period of
only a generation or two. However, if we look back into time we
can see a paleontological record displaying an ebb and flow of life,
great advances and great recessions, which, like the real ocean

tides, seem to be associated with natural phenomena that move as purposelessly as the moon and sun.

Fossils preserve this record of the waxing and waning of earthly life, primarily the life that lived around the fringes of the ancient seas, in marshes and lakes. These wispy echoes of life forms a billion years dead have been measured and described at great length in the journals of paleontology. Taken with the data of radioactive dating, a history of evolution emerges. Working with modern analytical techniques, such as protein analysis and comparative embryology, paleontologists draw the well-known "trees of life," showing branching upward of living things into the orders and families of plants and animals we know today. Some branches do not prosper, and that branch of the tree dies. Other branches appear weak and sickly during one geological age only to blossom forth with great vigor in the next. In general, the fossil record tells

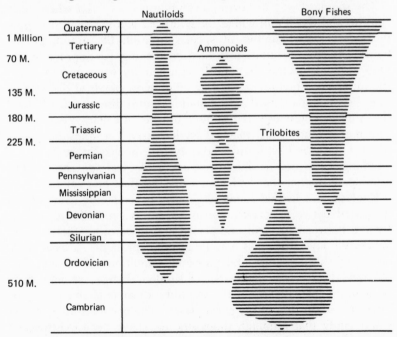

A typical evolutionary chart showing the ebb and flow of animal families as a function of geologic time. The width of the shaded areas is proportional to the number of families in existence. Note the extinction of trilobites after the Devonian period.

of an overall trend toward the higher, more complex forms of life. But it is not a steady progression; evolution seems to come in spurts, although the spurts may last millions of years.

Evolution is not measured by sheer abundance of life but rather by the rate at which new species, orders, and families are created. The Cretaceous period, for example, is always held up as the Age of Reptiles, when one form of life met with great success and roamed the earth in large numbers. We will never know how far the dinosaurs would have gone down the evolutionary trail; they were swept away by a changed environment that their gene reservoir could not cope with. As a result of this change in the world, whatever it might have been, dinosaurs were out and mammals were in. The same "housecleaning" can be observed in other groups of animals and plants at the boundaries between other geologic periods.

The standard way to portray evolution is on a chart where time runs up and down and where the widths of vertical bands are proportional to the number of species in each life grouping (see figure). Assuming evolution to be a good thing, we approve as the vertical bands on the chart swell with new life forms but frown at the sudden contraction and tailing off of many bands at the boundaries between geological ages. Should we also disapprove of bands that stay monotonously the same width because they seem to signify that evolution has reached an impasse? The question posed but not answered with any certainty in this chapter is: What, if anything, happened at the end of one geological age and the beginning of another that had such a drastic effect on evolution?

Geology tells us that something, possibly something cataclysmic, happened between periods, because there is generally an *unconformity* between rocks of one period and those of another; for example, the layers of rocks may be at different angles, a clear indication of shifting crust and changing conditions of erosion and sedimentation. These hallmarks of sudden change are important because one explanation of the throttling of some channels of evolution is "exhaustion of the germ plasm." The unconformities in the strata imply that although dinosaurs may have been overspecialized and possessed a gene reservoir inadequate to coping with environmental changes, they did not just fade out of the picture, they were pushed out violently and rather quickly on the geologic time scale.

We view the Four Horsemen of the Apocalypse with foreboding, and a forest fire is a terrible thing; yet life springs forth with renewed vigor after fire sweeps through a mature forest. Similarly, in evolution long periods of stability are followed by a crisis that pushes evolution another step up the ladder.

The fossil record is rather specific—at least for those organisms that died in just the right way and at just the right place, so that they were interred in sedimentary rocks that would one day be accessible to the questing paleontologist. Of the roughly 2,500 families of animals that have left their signatures on sedimentary rocks, only a third survive today. Many have dropped out of sight completely, while others have evolved into new families. Despite the high attrition rate, life on the average gains more species than it loses—more complex species in general.

A classic case of "extinction," as paleontologists are wont to call these sudden pinching offs in their bar graphs, involves the trilobites. Seen in great abundance in limestone and sandstone from the Cambrian period, these little primitive relatives of the lobster were almost wiped out by the end of the Devonian period.

A similar but less complete extinction transpired at the end of the Ordovician period, when half the fish families and 60 percent of the surviving and newly evolved tribolites met their dooms.

Near the end of the Permian, almost half of the known families of animals disappeared from the face of the earth. Many marine invertebrates were extinguished completely. Eighty percent of the reptile families and 75 percent of the amphibian families followed the same path as the Permian drew to a close. Thus, the sea has provided us with yet another "mystery" to explain.

No scientist denies these wholesale extinctions. The question is whether the extinctions occurred in a brief violent interlude that could be termed Catastrophic or whether there were slow Uniformitarian-type changes over millennia (rather than days or years) that gradually forced unadaptable species into sedimentary graveyards.

On the Catastrophism side of the argument we find our old friends Price, Donnelly, and Velikovsky. The stage is set by a quote from Price's *Evolutionary Geology and the New Catastrophism:*

Rocks belonging to all the various systems or formations give us fossils in such a state of preservation, and heaped together in such astonishing numbers, that we cannot resist the conviction that the majority of these deposits were formed in some sudden and not modern manner, catastrophic in nature.

Fossil deposits answering this description are found all over the world. The famous "Old Red Sandstone" from the Devonian period astonished the early geologists with its abundant fossil record. A record that on first sight seemed to be a tale of sudden, wholesale death. This kind of evidence, showing huge accumulations of animal remains, is akin to the piles of skeletons found in caves by geologists during the eighteenth century and attributed to the Flood (chapter 6).

In the Uniformitarian picture we envision dead fish sinking to the bottom, the carcasses rotting and being preyed upon, until all that the fossil record might show would be a few dispersed bones. Instead, we find huge schools of fish of the same species buried together, often facing the same direction. Pages from the fossil record confirm some aspects of Price's picture; but generally paleontologists do not hold with worldwide catastrophes. There were localized catastrophes, which created evidence with a Price flavor, but paleontologists believe the bulk of the fossils created during an "extinction" were deposited over many thousands of years. Again the time factor distinguishes Uniformitarianism from Catastrophism.

Fish kills are increasingly common these days. Every summer brings tales of windrows of dead fish stretching for miles along our shores. Many fish kills are the result of civilization's pollution, but others are natural; viz., those associated with the so-called Red Tide along the Gulf Coast caused by the release of toxins by rapidly proliferating phytoplankton. In other words some natural forces of extinction are still operating today and in a Uniformitarian way. The huge fossil deposits made up of species that had reached the ends of their respective evolutionary branches could have died this way. We need not invoke the holocaust of sudden celestial bombardment and crustal upheaval to explain the fossil evidence. If only our radioactive dating techniques were more precise, we could tell whether these rocky graveyards were filled overnight or over several millennia.

Whether an event is Uniformitarian or Catastrophic is a matter of interpretation and philosophical leaning. A Red Tide is certainly catastrophic to the huge rafts of dead fish that wash ashore and offend the nose; however, present-day natural fish kills are not worldwide and, because they are also the result of known forces, they are compatible with Uniformitarianism. No *localized* catastrophe—volcano, earthquake, flood, plague—is opposed by the Uniformitarian. He opposes only sudden, worldwide convulsions of nature, in particular those with supernatural overtones, such as the dunking of Atlantis by the gods or the opening of the Red Sea.

Of "catastrophes" acceptable to the Uniformitarians there is no shortage. Most of those listed in the table of chapter 1 are acceptable, if limited in geographic extent and stretched out in time. The exceptions are those celestial actions involving the moon and the other planets of the solar system. The Uniformitarians especially like forces of extinction that they now see in operation. Some acceptable causes of extinction are:

Radiation: Heavy doses of radiation could result if the earth's magnetic field collapsed; say, during an intense solar storm. The "magnetic bottle" we live in protects us from much of the particulate radiation arriving from the sun. However, calculations made by several scientists seem to show that the increase in dosage due to the loss of the field would not be sufficient to kill organisms directly, assuming the current levels of cosmic and solar radiation. For radiation to be a valid cause of catastrophe, colossal solar eruptions or nearby supernovas have to be postulated (as they have been by several scientists). Direct death by radiation is possible, and, in addition, over a few thousand years genetic damage might also cause extinction for susceptible species.

Pandemics: Populations of animals (and people) have been decimated within historical times by various diseases. One scientist, D. Newell, has proposed that the dinosaurs succumbed to pathogenic fungi. Usually some small part of a population survives any pandemic either because they somehow miss being exposed or have sufficient resistance conferred by their genes.

Poisons: Poisons of biological origin, such as the Red Tide toxins, are possible causes of extinction. The lack of biologically important trace elements, such as copper, during past geological ages has also been suggested as a cause for the withering and death of some species. In a sense we have a form of starvation here.

Climate changes: The great catastrophes of nature postulated by

Price, Velikovsky, et al., had to result in radical climate changes. The Uniformitarian can accept slowly developing changes in climate, such as those that brought on the Ice Ages, but he rejects any sudden tilting of the earth's axis and similar gyrations of the earth. Slow climatic changes can be explained by an increased "greenhouse effect," due to trapping of more solar heat by increased carbon dioxide in the atmosphere, and other reasonable cause-effect situations. Unfortunately for this theory, the fossil plant record, which is considered a good key to ancient climates, shows only minor changes occurring at the closing of the Permian, Triassic, and Cretaceous periods, when so many animal families perished. But, perhaps the endurance of plants from one age to another is an indirect clue—plants are much less susceptible to radiation damage than animals.

Geologic revolutions: A geologic revolution is a Uniformitarian way of saying "slow catastrophe." It should be taken in the sense of widespread changes in the surface of the earth on account of internal readjustments, due perhaps to expansion of the earth, but certainly not, according to Uniformitarianism, to astronomical cataclysms. All geologists from Cuvier on have been impressed by the fact that many erosional unconformities, indicating that something has happened to change drastically conditions of sedimentation, occur hand in hand with the last fossil records of many species. Again, it seems a matter of semantics and time scale. Something happened at these breaks in the erosion pattern, something that may also have killed off large fractions of the earth's animal population; but we really do not know whether it happened quickly or slowly or what stimulated it in the first place. The fossil record has amnesia during the periods when the unconformities were created.

Man: One great exterminator we know a lot about is man himself. Man with his guns and pesticides, with his farmlands and cities, has been directly responsible for the extinction or near-extinction of some 450 easily observable species within historical times. No counts of insect or microorganism extinctions can be made. The fossil record now being written may relate these facts to some paleontologist of the future, but we cannot blame man for what happened in the Age of Fishes or the Age of Reptiles. The recent Pleistocene period, however, is another story, for those were the times when man became an efficient big game hunter. The fact that at least two hundred genera of large animals came to the end of evolution's road during this period, has encouraged some scientists to place the blame on man rather than climate or some other physical force, although man has been a relatively rare animal until recently.

How can scientists blame any specific cause for extinctions that transpired under the Permian sun or while death gripped the planet from pole to pole at the end of the Cretaceous? "Guilt by association" is the only way we know. If we could correlate in time evidence of some causative agent with the extinctions incised in the layers of sedimentary rock, we would have a case.

Looking for causal connections between climate, radiation level, rate of evolution, astronomical phenomena, sea level, and the many other parameters that have varied over the last 4.5 billion years is more of an art than a science. One recalls the sunspot cycle craze during the 1930's and 1940's, when everything from the stock market to the weather was slaved to the master eleven-year solar cycle. Evolutionary "pulsations" are not as regular as the sunspot cycle (itself rather variable). It is easy to confuse cause and effect and still easier to associate unrelated parameters when the data are old and so imprecise. Nevertheless, some of the correlations that have been found are remarkable and probably testify to the occurrence of rather spectacular, far-reaching changes in the earth's atmosphere, lithosphere, hydrosphere, and, of course, biosphere. Putting it more directly, these "spheres" seem to have been stimulated repeatedly and simultaneously by mysterious causative forces.

With only the fossil evidence at hand, the purveyors of Catastrophism postulated sundry celestial encounters as the preceding chapters have described. As the reader will recall, limited Catastrophism has enjoyed a renaissance due to the confirmation of continental drift, sea-floor spreading, magnetic reversals, and other signs that the earth's past was not exactly serene. Two leading figures in "neocatastrophism" are Robert J. Uffen, a professor at the University of Western Ontario, and Bruce C. Heezen, a geologist at Columbia's Lamont Geological Observatory. Heezen was (with Marie Tharp) a discoverer of the mid-Atlantic Rift (chapter 3). Uffen published some of his first speculations about the role of radiation in the evolution of life in *Nature*, in April 1963. During the next few years Uffen, Heezen, Billy Glass, John F. Simpson, and several other investigators put together a jigsaw puzzle of magnetic, deep-sea, and fossil evidence that correlated the ebb and flow of life with our planet's troubled past. It is only fitting, however, that we begin with a quotation from *The Surface History of the Earth* by the famous Irish physicist, John Joly, who said in 1925 in support of radiation Catastrophism: ". . . had our world been montonously free from change, save diurnal and seasonal ones, much of the diversity of animal and vegetable life would never have been."

Uffen's 1963 thesis was that variations in the earth's magnetic field strength, controlled by internal molten convection cells, in effect raise and lower the shields protecting the earth's denizens from cosmic rays and intense solar particulate radiation. In 1963 no convincing correlations connecting life with magnetism were available. The first compelling evidence supporting Uffen came from deep-sea cores in 1964 and 1965. The cores showed biological extinctions correlated in time with reversals of the magnetic field. A paper published by N. D. Opdyke and Billy Glass in the October 21, 1966, issue of *Science* stated: "The coincidence or near-coincidence of faunal changes with reversals in these cores suggests a causal relation." The authors went on to conjecture that the increased radiation levels would have upped the mutation rate and "strongly affected the evolutionary process." These theories do not depend upon outright radiation death—a rather Catastrophic occurrence that is hard to support physically on the basis of currently measured radiation levels out in space. Instead, merely a strong increase in radiation level is postulated rather than the many orders of magnitude necessary for radiation death. A strong increase (two to ten times present levels) might trigger so many mutations that sensitive species could not maintain themselves, particularly if the environment also changed.

The grandest correlation of all came with a 1965 paper entitled "Evolutionary Pulsations and Geomagnetic Polarity" by John F. Simpson, a scientist at the University of Akron, in the *Bulletin of the Geological Society of America*. Simpson compared extinctions, new animal families, magnetic polarity, orogenic activity, and other geological factors as a function of geologic time (see figure). The correlation of "evolutionary pulsations" with *changes* in the polarity of the terrestrial magnetic field is quite striking. When the polarities of samples shift from plus to minus, or vice versa, the number of animal families often increases. There are also some

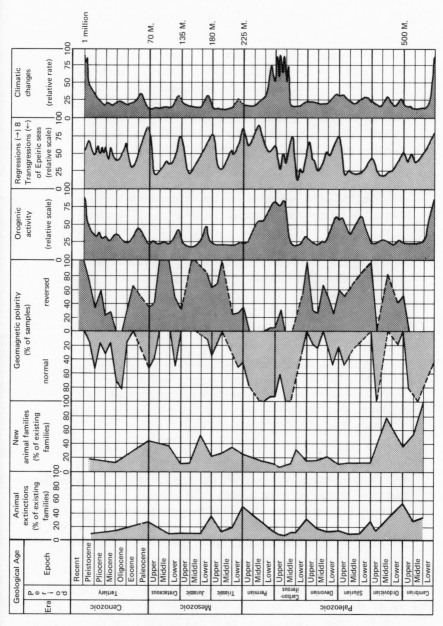

Correlation of various parameters associated in some way with the rate of evolution. (Reproduced with permission from J. F. Simpson, *Bulletin of the Geological Society of America, 77,* Feb. 1966, p. 197)

connections between orogeny (mountain building) and the subsequent abundance of species at various interfaces between geological periods. Nor can one say that our planet's climate and oceans are completely divorced from the biological pulsations. And which, pray tell, is the cause, and which the effect? The primary cause may not even be inscribed on the chart.

Further substantiation of the connection between the ups and downs of the terrestrial magnetic field and those of life have come from many scientific quarters. Indeed, the reversals and the waxing and waning of our protective magnetic field seem to be correlated with the Ice Ages, sea level, cycles of mountain building, and all manner of earthly tribulations. It seems as if we have another cyclic time marker like the sunspots for important terrestrial events.

Such a bold hypothesis could not go unchallenged. The main objection offered is that the current radiation levels in space are not high enough to cause significant genetic damage even if allowed to bathe the earth without the protecting magnetic field. C. J. Waddington voiced this objection in the November 17, 1967, issue of *Science:*

> The hypothesis that the additional energetic particle radiation allowed to fall on the earth when the geomagnetic field is reversed is the causative agency for population changes thus appears untenable unless it is assumed that these periods are associated with greatly increased particle radiation from some external source. *

More succinctly, just reversing or lowering the earth's magnetic field does not seem to be enough to account for the widespread evolutionary changes observed in the fossil record. The important hint here is that perhaps something else besides a change in the magnetic field is involved, possibly increased outpourings of radiation from the sun. It may have been that radiation was not involved in extinction at all and that it, like the magnetic field pulsations, was an effect rather than the primary cause.

Billy Glass's discovery of tektites in the Antarctic sea-bottom cores added another factor to the equation. There, in the same sediments that causally connected magnetic and biological pulsations, were tiny fused bits of glass that had entered the earth's at-

---

* Copyright 1967 by the American Association for the Advancement of Science.

mosphere from outer space at high velocities (chapter 3). According to current theories, the tektites could have come from a comet, from rocky debris blasted from the moon's surface by a meteorite, or from debris from a similar impact on the earth that arced over in a suborbital trajectory seeding wide areas of the earth with the tiny glass beads. Whatever the source of the tektites, they were coincident with magnetic reversals and evolutionary pulsations. This discovery threw a distinctly Catastrophic light on the whole business of cause and effect in evolution. It was the comet possibility that gave John Lear the title for his controversial 1967 *Saturday Review* article, "Were Comets the Midwives at the Birth of Man?" If ever a title had Catastrophic overtone calculated to annoy Uniformitarians, this was it.

From the work of Uffen, Heezen, Glass, and Simpson, we have the essentials of an integrated Catastrophic hypothesis, although no one has called it such directly. A heavenly body wounds the earth or moon, stirs up the tektites, reverses or reduces the earth's magnetic field, allows radiation to bathe the unprotected surface, and quite likely stimulates earthquakes and other terrestrial cataclysms. (Earthquakes apparently can be brought on by the earth's wobble.) Donnelly or Velikovsky could not have proposed anything to suit their tastes better. It is for this reason that the label "neocatastrophic" is most apt for such hypotheses.

Astronomical Catastrophism is still frowned upon in scientific circles as too redolent of fire and brimstone and other suspicious deep dark memories of the race's past. So, let us look for other prime causes. A hypothesis cautiously ventured by K. D. Terry and W. H. Tucker in 1968 depended upon radiation from nearby supernovas to extinguish species upon the earth. This thesis was soon challenged, on a probabilistic basis, by Howard Laster, at the University of Maryland. Laster stated that the radiations from supernovas would be too weak and too spread out in the time dimension, unless they were very close to the earth, and that supernovas close enough to the earth to do any biological damage were too infrequent to account for all of the observed evolutionary pulsations. However, the Apollo astronauts added a new possibility when they discovered that many small lunar craters showed glazing at their centers. The implication is that flare-ups of the sun melted patches of lunar soil. The earth was undoubtedly exposed at the same time, but the atmosphere absorbed much of the heat. Solar cosmic rays

EFFECTS	SOURCE OF EVIDENCE
Magnetic reversals	Deep-sea cores
Tektite deposits	Deep-sea cores
Life extinctions	Fossil record
Unconformities	Visual inspection of strata
Climate changes	Fossil record
Sea-level changes	Sedimentary deposits

accompanying the flare-ups could have been biologically important, however.

For those absolutely opposed to celestial intervention, we have the theory of cyclic orogeny proposed by John Joly. In essence, as discussed in chapter 4, Joly postulated that cycles of mountain building and life were linked causally to upwellings of hot magma driven by the heat of radioactivity. Obviously, orogeny would change climatic patterns, nuclear radiation from the radioactively driven convection cells might also affect the course of evolution, and the varying currents of electrically conducting subterranean magma would enhance and occasionally reverse the earth's field. The only variable that Joly did not include in his equation was the existence of tektites in deep-sea cores. Otherwise, he would have had a neat, tidy hypothesis that was independent of celestial activity.

So many possible causes and effects are mixed up in the preceding paragraphs that a table to sort them out is in order.

The table has Catastrophic predilictions. It may be, of course, that none of these events are causally related and that each has its own Uniformitarian cause. Accepting the possibility of a common cause, we see that intense solar storms do not have the wide range of effects attributable to orogeny and celestial Catastrophism. Comparing Joly's radioactive cycles of orogeny with the comet/meteor hypothesis, the tektite fields associated with faunal extinctions and magnetic reversals seem to tip the balance in favor of some form of celestial Catastrophism. Unquestionably, cyclic orogeny would be preferred by Uniformitarians because Joly's radioactive engine operates on physical principles in evidence today and does not depend upon chance visitations from outer space. But as we observe the pockmarked surfaces of Mars and the moon

INTENSE SOLAR STORMS	CYCLIC OROGENY (MOUNTAIN BUILDING)	METEOR, COMET, ETC.
X	X	X
		X
X	X	X
	X	X
	X	X
	X	X

through the eyes of astronauts and remotely controlled machines, who can deny the long-term effects of celestial bombardment? Further, many fresh craters with sharply defined rims have been seen close-up on the moon, indicating that powerful projectiles still ply the not-so-empty space near the earth.

Real proof of the reality of Catastrophism will come only when some comet or small planetoid impacts the earth, setting off in various degrees the consequences listed in the table. We very nearly had proof in 1908 when the famous Siberian "meteorite" (possibly a piece of antimatter) leveled whole forests in that remote region and jiggled barographs around the world. Meanwhile, science generally sets a Uniformitarian course because only in Uniformitarian terms does it have a predictable world that it can explain by current physical processes. Yet Catastrophism is growing in popularity as more and more evidence, from deep-sea research in particular, accumulates, suggesting violent, random events in the earth's past.

What is the next step in the evolution of the earth and the life upon it? Or will there be a next step? Is it too presumptuous of man to think that he is the last step in this multibillion-year-long drama? Given the present data that science has gleaned from that tiny part of the universe observable to it, there is no objective way of answering these questions. This leaves us free to speculate, while continuing the search for additional data.

George Gaylord Simpson, one of the great modern interpreters of evolution, stated in his book, *The Meaning of Evolution:* "New major sorts of organisms have arisen, as a rule, not as more effective followers of ways of life already occupied but as groups extending into and eventually filling new ways." This sentence is, of course, consistent with the body of this chapter where it was pointed out that life usually flowered abundantly during and shortly after changes in geological periods when the environment was changing rapidly. The changing environment, then, screened out those species that could not prosper under the new conditions and, at the same time, created ecological niches for whole new families of animals. All life was not extinguished at the geologic interfaces. Many of the most primitive forms of animals are still with us almost unaltered today; and many other families survive because they took new tacks along evolution's way. In general, the record shows an overall enhancement of life over the billions of years. Would life have prospered so well without the spice of environmental change? This question echoes John Lear's *Saturday Review* title, "Were Comets the Midwives at the Birth of Man?" Catastrophists and many Uniformitarians are beginning to think that man would never have come to be without the stimuli of comets, meteorites, and other heavenly utensils to stir the fecund soup of life on earth. Naturally, no one can yet substantiate or disprove such speculations; but they are intriguing.

Where will man go from here? No interlopers are on a collision course with the earth according to the astronomers' telescopes, and new cycles of orogeny do not seem to be building beneath our feet. But it is doubtful that evolution has run its course. Mankind would like to conceive of the next step in evolution as being a sociological one in which he learns to live with himself. Catastrophism may take the matter out of his hands. Not astronomical or geological Catastrophism but man-made Catastrophism. The use of this planet's atmosphere and hydrosphere as civilization's garbage dump may ultimately trigger planetwide climate changes or upset some of nature's delicate balances. Scientists have been concerned for decades that the carbon dioxide we pour into the air will trap solar heat, raise the earth's average temperature, and melt the polar ice cap. Such catastrophes would have drastic effects on life. They may not come overnight, with dawn finding the great cities asphyxiated by their own effluents; but the quality of life and the average level of health might deteriorate steadily as populations soar and smokestacks belch.

Man and his machines have already affected the earth's environment and inventory of life irrevocably. Some species have been extinguished altogether; others, especially insects and microorganisms have responded to man's changes (DDT, penicillin, etc.) by evolving in ways to better insure their survival. Undoubtedly, man is also evolving due to feedback through the environment. Just how much and in which direction, we do not know.

The scientist takes a bright view of the universe: he sees no imminent catastrophes; he sees instead an opportunity to better understand the way the universe operates and with that knowledge build a Utopia. If science can control the environment and if man can control himself, a high rung of the ladder has been won after four billion years. How will life evolve to reach the next higher step?

READING LIST

COX, A.: "Geomagnetic Reversals," *Science,* Jan. 17, 1969.

GLASS, B., and HEEZEN, B. C.: "Tektites and Geomagnetic Reversals," *Nature,* Apr. 22, 1967.

HEEZEN, B. C.: "Tektites and Geomagnetic Reversals," *Scientific American,* July 1967.

LEAR, J.: "Is Magnetism the Key to Life's Origin on Earth?" *Saturday Review,* July 6, 1963.

—— "Were Comets the Midwives at the Birth of Man?" *Saturday Review,* May 6, 1967.

MARTIN, P. S., and WRIGHT, H. E., eds.: *Pleistocene Extinctions, the Search for a Cause,* Yale University Press, New Haven, 1967.

NEWELL, N. C.: "Crises in the History of Life," *Scientific American,* Feb. 1963.

SIMPSON, G. G.: *The Meaning of Evolution,* Yale University Press, New Haven, 1967.

SIMPSON, J. F.: "Evolutionary Pulsations and Geomagnetic Polarity," *Bulletin of the Geological Society of America,* Feb. 1966.

10

LIVING FOSSILS
AND SEA MONSTERS

Lopping off the firehose-size arms of the kracken with axes, Captain Nemo and the crew of the *Nautilus* finally escaped this monster from the ocean deeps. Jules Verne had ample basis for his tale because ever since men have taken to the sea, they come back with descriptions of huge sea beasts of ferocious mein, sinuous creatures large enough to engulf whole ships. Bolstering such stories were occasional carcasses washed ashore—unidentifiable, mysterious, and malodorous. Whaling ships returned to port telling of sperm whales they had captured—no little animals themselves—which bore huge circular scars of their battles with leviathans of the abyss.

The whole "sea monster" or "sea serpent" business is suffused with exaggeration, hoax, scientific incredulity, poor data (from a scientific standpoint), and some germs of truth. It is a philosophical brother to Catastrophism, for what could be more romantic than dreams of the dim past when great forces beyond our control seized the earth, when continents sank and were wrenched apart, and monsters swam in the seas?

It may be that sea monsters (if they really exist) are holdovers from past geological ages; that is, "living fossils," relics that somehow escaped extinction. However, size alone need not be the sole criterion. Many shellfish and small sea animals living today have survived evolutionary and environmental pressures for hundreds of millions of years. We want to inquire why such creatures still survive, because their hardiness may help us gauge the nature and magnitudes of the forces that extinguished so many species in past aeons.

The whole sea-monster business is highly pertinent to our attempts to understand the earth's past by studying the sea for an entirely different kind of reason. The stories about strange sea creatures are on the periphery of science, like sightings of drowned cities from aircraft (chapter 7) or the legends of fire and brimstone falling from the sky. Testimonial, unverifiable data could play a major role in the frontier lands of science, if we only knew how to use them properly. The story of sea monsters is more than a collection of curious sightings; it is a case history of how accumulated "soft" data can be used to help convince the scientific community of the reality of things that cannot be deliberately verified.

Why should the possible existence of a large, undescribed sea creature be the butt of scientists' scorn? It cannot be size, bizarreness, or habits; for in their questing nets oceanographic expeditions have brought up all manner of weird creatures, and the blue whale puts the dinosaurs to shame in the matter of volume. Sea serpents have three things working against them:

(1) Testimonial evidence of their existence comes almost exclusively from laymen, not from scientists.
(2) No bona fide, well-authenticated specimens exist in the museums.
(3) Sea serpents are associated with legend, myth, and romance, which science feels obliged to expurgate.

Sea serpents are not the only natural phenomena facing these same roadblocks on the road to respectability. The abominable snowman is a notorious example still outside the field of science. Stones falling from the sky were also once placed in this category; so was the pigmy hippopotamus. With due consideration of the fact that science must maintain high standards to repel cranks and perpetrators of hoaxes, sea serpents seem to have been treated unfairly. Freud might have ventured that some unconscious prejudice is still carried over from the Garden of Eden story and the almost innate human fear of snakes. More seriously, though, science seems on the verge of accepting sea serpents as legitimate denizens of the ocean depths and, though many may flinch at the thought, Loch

Ness, too. If the barriers are lowered to sea serpents, much of the credit will have to go to one man: Bernard Heuvelmans, a professional zoologist and author of the incipient classic *In the Wake of the Sea Serpents.*

This 645-page treatise on sea serpents, their lore, their sightings, and the centuries of controversy swirling around them, must have taken courage to write. Professional destruction is sometimes the lot of those who stray too far into the so-called "lunatic" fringe areas. Nevertheless, Heuvelmans has apparently won. He has now written two books on "monsters" and other strange elusive animals once ridiculed by science. The predecessor of *In the Wake of the Sea Serpents* was *On the Track of Unknown Animals.* Both have received favorable reviews in many journals. Part of their acceptance derives from the thorough, discriminating jobs that Heuvelmans has done; partly because he is recognized as knowledgeable in zoology, and partly because the idea that sea serpents still exist today is appealing, almost hypnotic. Mysteries are fun if no personal risk is involved.

In surveying other fringe-area classics introduced in this book, we find that the books of Heuvelmans are rather like Wegener's *The Origin of Continents and Oceans.* As was Wegener, Heuvelmans is thoughtful and objective; he throws out questionable observations and any data tainted by the hint of hoax. Both Wegener and Heuvelmans were, of course, trained scientists. Donnelly's books and those of Velikovsky, regardless of any germs of truth they may contain, were much less carefully done; they were not endowed with obvious objectivity; they leapt to conclusions. Both Donnelly and Velikovsky displayed tremendous energy in putting their books together, but their standards were low and they were blinded by their theses when it came to accepting or rejecting data. This is the major reason for the considerable, but not yet complete, success of Heuvelmans and the complete rejection of Donnelly and Velikovsky by science.

Heuvelmans' data are almost all of the testimonial type. Someone, an experienced seaman in most instances, sees a large unusual animal. Being used to the ways of the sea, he knows what's usual and what's not; though he can be fooled sometimes by floating seaweed and rare, but scientifically acceptable, creatures from the deep. The snakelike oarfish causes cases of confused identity. Then, again, sailors sometimes succumb to the lure of the hoax. Still, when all the questionable data are trimmed away, there is left a substantial body of information that cannot be dismissed. Something must be done with these data; they cannot be brushed under the scientific rug forever.

Some sea-serpent sightings are sharp, clear, consistent, and unembellished by spectacular claims. Indeed, Heuvelmans asserts that hoaxes and false reports can be separated from the honest claims with surprising ease. Many sea serpents were seen by several people at the same time. The cross-checking of stories, in good lawyer fashion, usually reveals any hoax or lie. Cross-checking also reveals that people are usually notoriously poor observers. Nevertheless, Dr. Frederic A. Lucas, director of the American Museum of Natural History for twenty years, once stated that the sworn evidence attesting to the reality of sea serpents would be more than adequate in any court of law.

Testimonial data is usually adequate when the viewer is a well-trained scientist. For example, there are several species of cetaceans (whales, porpoises, etc.) that are generally accepted by scientists, even though they have never been able to examine a carcass. Several marine zoologists through the years have seen a cetacean with two dorsal fins. This species has even been given a scientific name (*Delphinus rhinoceros*). Such acceptance holds for many other sea creatures, for birds, and land animals. Yet sea captains and their crews, all familiar with the sea, are discounted when they tell of sea serpents. However, the weight of the data assembled by Heuvelmans has begun to tip the balance in the positive direction. He reports on 587 sea-serpent sightings from 1639 to 1964, many by several capable observers at the same time. Either there is something to sea serpents or a lot of respectable people are either liars or simply mistaken.

Most sea monsters are not really serpentine. Many, in fact, seem to be large mammals rather than reptiles. However, one of the greatest sea monsters of all is a cephalopod, a mollusk that has outgrown his shell many times over, the giant squid, or kracken. Sometimes more than fifty feet long, the kracken, when finally recognized by science, became the largest invertebrate.

Like Atlantis, the kracken swims into our ken in the writings of the ancient Greeks. Scylla, in the *Odyssey*, has many attributes of the giant squid. Even more astounding accounts of this huge sea creature came from Norwegian sailors in the eighteenth century.

When a huge kracken surfaced, they said, it would form an island a mile and a half in circumference. Sailors mistaking the animal for land sometimes would row over to it, build a fire, and settle down for the night, only to be left thrashing in the sea when the kracken got tired of being a hearth. Sailors do exaggerate, one must admit. There is also a Norwegian legend that tells us that only two kracken were created at the beginning of the world and that they will surface and die when the end of the world is imminent. The kracken stories are not unlike those of Atlantis and Leviathan in the Bible; they have grandeur and tell of forces beyond those under man's control. Yet real kracken do exist; not as huge and terror-inspiring as myth and legend would have them; but big enough to show us how myth and legend got started in the first place.

Reality was hard to prove against the colorful background of legend and overembellishment. When Scandinavian fishermen told of huge squid in the middle 1800's, they were held to be under the influence of a new brand of rum that had been introduced in the coastal areas. No reputable scientists believed that giant squid existed. Then, a few weeks before Christmas 1861 the French corvette *Alecton* made port with a wild story about a squid almost twenty-four feet long; it was too big to be believed. The *Alecton*'s captain, Lieutenant Bouyer, told how he had maneuvered his ship alongside an object floating in the sea to find a bright brick-red body, fifteen to eighteen feet long, with tentacles five or six feet long, and plate-sized eyes that stared back glassily. The monster seemed wounded somehow, and the *Alecton* tried to finish the job with harpoons. Nothing had much effect on the soft body of the animal. The crew finally got a rope around the big "tail" (if such a weird animal can be said to have a front and back) and tried to haul it aboard. The kracken's flesh was so soft that the rope sawed right through the tail, and the bulk of the animal slid beneath the waves for good, leaving the *Alecton* with only a "tail." Soon, even the tail had to be jettisoned as it ripened under the hot sun. The men of the *Alecton* could only tell what happened; they had no physical proof at all.

The *Alecton*'s report was discussed before the prestigious French Academy of Sciences. Because Lieutenant Bouyer and his men seemed reputable, the only conclusion the academy could come to was that they had been the victims of *mass hallucination!* How susceptible laymen are to this malady. Arthur Mangin, of the academy, concluded that the *Alecton's* monster was doubtless a large sea plant—a case of mistaken identity. He went on further:

> These considerations, and others as well, which a little reflection will no doubt suggest to the reader as well as ourselves, will suffice, I think, to persuade the wise and especially the man of science, not to admit into the catalogue those stories which mention extraordinary creatures like the sea-serpent and the giant squid, the existence of which would be in some sort of contradiction of the great laws of harmony and equilibrium which have sovereign rule over living nature as well as senseless and inert matter.

The French academy believed in a harmonious, Uniformitarian world, where kracken and sea serpents were discordant chords that must be suppressed.

Unperturbed by Mangin, Jules Verne converted the *Alecton* story into the kracken's attack on the *Nautilus* in *Twenty Thousand Leagues Under the Sea*.

As if in rejoinder to the conclusions of the academy and in support of Verne's fiction, nature quickly provided more kracken for men of learning to digest. Rather indigestible bits they were, too, for huge rotting carcasses were found stranded on beaches around the northern Atlantic, particularly Newfoundland and Scandinavia. Specimens eventually found their ways into museums and the hands of scientists. The largest kracken report fully accepted by science is the so-called Thimble Tickle squid. On November 2, 1878, during a period when kracken seemed to be committing mass suicide upon the beaches of Newfoundland, three fishermen from Thimble Tickle in a boat just offshore noticed a huge animal trying to escape being stranded by the tide. In a burst of courage, they snagged the fleshy part of the squid with a barbed grapnel and tied it with a stout rope to a tree on shore. When the tide receded, the kracken's body was measured as twenty feet long plus another thirty-five feet for the tentacles. We now are assured that kracken more than fifty feet long dwell in the deeps, despite the judgments of the French academy and the wild myths surrounding these fantastic creatures.

Were the stories of island-size kracken really so wild? The Thimble Tickle squid had suckers about four inches across. Yet whalers tell of finding sperm whales, who seem to dote on meals of kracken, bearing sucker marks nearly twenty inches across, five times the size of the Thimble Tickle squid. Thus, we have suspect evidence for kracken 5 × 50 = 250 feet long. Who really knows what monsters patrol the deep abyss where we still grope with crude tools?

Animals stranger than the kracken and sea serpents live in the ocean; they are just smaller. A confirmed sea serpent would just be one more new species to add to the lists of mammals and reptiles already many thousands of names long. Nonetheless, it is the sea serpent, not a lowly sea worm or crustacean, that captures our fancy.

Aristotle took note of sea serpents as he did of just about everything in the natural world. They attack ships, he said in *Historia Animalium,* throwing themselves on a trireme and capsizing it. No doubt some sea serpents could upset a small ship. Whales sometimes turned upon their tormentors, too. Other legends and myths are even more bizarre. Like the kracken, sea serpents were evil omens. Sailors in the Middle Ages feared them as much as they dreaded the great maelstroms and that terrifying rim of the world where the ocean waters cascaded into infinity. Many American Indian tribes have legends of gigantic water serpents; some in deep, freshwater lakes (like the Scot's Loch Ness!) and some that inhabited salt water.

To begin an inquiry, let us look at a few of the more voluble sea-serpent champions. In historical order we have:

Professor Antoon C. Oudemans, director of a zoological park in Holland and a member of the Royal Dutch Zoological Society. In 1862 Oudemans published his comprehensive *The Great Sea Serpent,* which in some 592 pages dealt with two hundred sea-serpent sightings.

Lieutenant-Commander Rupert T. Gould, of the British Royal Navy. An amateur zoologist, Gould was also an excellent mathematician. His 1930 book *The Case for the Sea Serpent* was extensive, thorough, and a model of scientific rigor. With his nautical knowledge, Gould checked logbooks, naval archives, everything he could, to make certain that the cases he studied were truthful. (In passing, we note that Commander Gould also authored the books *Oddities* [1928] and *Enigmas* [1929], in which he encompasses many other unusual phenomena encountered on this planet down through recorded time.)

Tim Dinsdale, the enthusiastic chronicler of *Nessie* (the Loch Ness monster to the uninitiated). Dinsdale is an aerospace engineer with a penchant for the unusual, monster-hunting in particular. His two books on sea serpents, *The Leviathans* and *Loch Ness Monster,* are carefully done though perhaps more "pro" than a professional zoologist would have made them.

Bernard Heuvelmans, met earlier in this chapter, and the author of the latest synthesis, *In the Wake of the Sea Serpents.*

The pro-serpent contributions of Maurice Burton, A. Hyatt Verrill, Willy Ley, Richard Carrington, and Ivan T. Sanderson also must be mentioned. These latter gentlemen are authors of books that deal with sea serpents but only in part. They are all noted for their popular works on zoology.

All in all, the list of protagonists is longer and boasts a more professional sheen than, say, the list of supporters of Atlantis. This may signify that sea serpents are more likely to be real.

All sea-serpent treatises follow the same path. It is indeed difficult to imagine any other road to truth in this case: testimonies are collected and analyzed. Anybody can do these things, and the prospects attract many amateurs. In the case of the Loch Ness monster, however, Dinsdale and other sea-serpent watchers actively patrol the loch with cameras. In an attempt to transcend testimonial evidence sonar surveys have also been made. Unlike Nessie, most sea serpents do not have captive audiences and must depend upon chance encounters to receive publicity.

In the sea serpent vignettes that follow, the reader should compare the type and quality of data employed to establish their existence with those data that are claimed to support the existence of Atlantis and past terrestrial Catastrophism. Myth and legend are involved in all. In the cases of the rare, unbelieved sightings of ephemeral islands, ruins of cities beneath the sea, and even occult contacts with past inhabitants of Atlantis, we again have testimonies from witnesses. But how reliable are the witnesses?

The testimonial evidence is almost endless. Heuvelmans covers 587 cases; and we may be sure that his book will stimulate many new ones, both real and imaginary. His book will also unlock many sightings formerly withheld for fear of ridicule. A quartet of the more famous sightings should suffice here.

In 1817 a sea serpent "flap" occurred in Massachusetts Bay. For two weeks during August many reports accumulated of a strange marine animal in Gloucester harbor whose length approached one hundred feet. The sightings were so numerous that the Linnaean Society of Boston set up a committee to carry the matter through to a scientific end. Hundreds of people saw the monster, some from

as close as twenty feet. There can be no doubt that something big
was sporting in Gloucester harbor that August.

The Linnaean Society took only written testimony, and each
contributor had to answer twenty-five questions prepared in advance. Shipmaster Solomon Allen testified that the monster was:

> . . . between eighty and ninety feet in length, and about the size of a
> half-barrel, apparently having joints from his head to his tail. I was
> about one hundred and fifty yards from him, when I judged him to be
> the size of a half-barrel. His head formed something like the head of
> the rattlesnake, but nearly as large as the head of a horse. When he
> moved on the surface of the water, his motion was slow, at times play-
> ing about in circles, and sometimes moving nearly straight forward.
> When he disappeared, he sunk apparently directly down, and would
> next appear at two hundred yards from where he disappeared, in two
> minutes. His color was dark brown, and I did not discover any spots
> upon him.

A clear, concise, unembellished report from a reputable witness.
Other affidavits supported Allen.

One feature of the Gloucester sea serpent that others mentioned
was the presence of humps or bunches on the animal's back. According to Willy Ley's *Exotic Zoology* the humps helped convert
the inquiry of the Linnaean Society from a masterful scientific
coup into an object of ridicule. It seems that the residents of
Gloucester figured that the serpent was visiting their shores for the
purpose of laying eggs. When a boy later found a three-foot-long
humped snake—a deformed black snake, in actuality—not only
the local residents but the Linnaean Society declared positively
that they had a bona fide specimen of a young sea serpent. This
mistake threw the whole Gloucester sea-serpent case out of scientific court and made the barrier erected against sea-serpent stories
a few feet higher.

Another famous story originated with the H.M.S. *Daedalus* on
August 6, 1848, in the South Atlantic, three hundred miles off the
coast of Africa. Captain Peter M'Quhae submitted to the Admiralty
a detailed report with a sketch of the creature he and his crew had
seen. Like all hard-core sea-serpent reports, the language was
sober and businesslike.

> . . . the object . . . was discovered to be an enormous serpent, with
> head and shoulders kept about four feet constantly above the surface
> of the sea, and as nearly as we could approximate by comparing it

with the length of what our main-topsail yard would show in the water, there was at the very least 60 feet of the animal . . . .

M'Quhae went on to describe the creature's head as being snake-like. Its neck was about fifteen or sixteen inches in diameter behind the head and something like a mane along its back.

Scientists rose to attack this outrageous sighting. Sir Richard Owen, a professor at the Royal College of Surgeons, said that the *Daedalus* serpent was merely an exaggerated sea lion or sea elephant, in other words a big mammal made bigger by unscientific observers. (Owen was soon to lead an attack on Darwin's theory of evolution.)

Sea serpents seem to like their privacy. Although the average number of good sightings per year hovers around a half dozen, it is possible that many go unseen because the propeller noise of modern ships may drive these creatures away from the shipping lanes into the less-frequented parts of the oceans. During wars, however, when more ships are at sea and surveillance more intense, they cannot hide quite so easily. One unusual account came from Georg Günther Freiherr von Forstner, the captain of the German submarine *U-28* during World War I. Stimulated by the Loch Ness monster headlines in the 1930's, he belatedly told how, on July 30, 1915, his U-boat sank the six-hundred-foot British steamer *Iberian*. About twenty-five seconds after the steamer plunged aft-first into the North Atlantic, the whole ship exploded. Blown clear out of the water was a crocodilelike monster, about sixty feet long, with four webbed feet and a long tail tapering to a point. It was gone in a few seconds. Additional sea monsters of this same general construction have been sighted by others.

It would be almost sacrilegious to leave the subject of sea serpents without saying something about the world-renowned Loch Ness monster. Hemmed in by splendid mountains, the waters of Loch Ness fill a long narrow wrinkle in the earth's crust about thirty miles long, terminating on the northeast at Inverness. No one knows just who saw Nessie first; suffice that it was long enough ago (at least A.D. 800) so that either several generations of monsters have flourished or possibly new members of the species have migrated in from the nearby Atlantic. Indeed, some reports tell of the monster being seen out of the water, and there have been monster sightings submitted from Loch Ness southwestward through other lochs to the Sea of the Hebrides. But Loch Ness is where the action is. At least two amateur societies have been formed, in true British fashion, to track down the monster. The Anglo-American Loch Ness Phenomena Investigation Bureau, Limited, obviously a nonprofit company, is one of these amateur organizations that has helped keep the loch under close surveillance in recent years.

The five thousand or so sightings claimed for Nessie over the centuries are similar in character to those of the saltwater sea monsters mentioned above. Of course, Nessie may be a true saltwater sea monster, whose descendants have survived in the loch for fifty to seventy centuries after being trapped by some catastrophic upheaval of the loch country. Many cold, deep lakes around the world have their own monster legends, but none as persistent as that of Loch Ness. The Loch Ness situation is especially important because someone can and is doing something about it. One cannot watch the oceans of the world, but a narrow lake thirty miles long is within reason. Amateur camera patrols have been maintaining vigilance over large portions of the loch during the summer months, with the help of a twenty-thousand-dollar three-year grant from the Field Enterprises Education Corporation, publishers of *World Book Encyclopedia*. Film has been shot of wakes purportedly left by the monster(s). Analysis of the footage by the Royal Air Force confirms the existence of a large submerged object in the lake. So far, though, the monster itself has not been captured conclusively on film, although there exist rather fuzzy, dubious shots of something poking above the water.

Tim Dinsdale, a long-time Nessie watcher, has summarized about one hundred sightings made since 1933. Nessie has a foot-thick neck about five feet long. The rest of the animal is always partly submerged, but it is estimated at between thirty and seventy feet. Most zoologists who deign to study the problem now believe that if a large animal swims Loch Ness it is probably a large mammal with front and rear flippers—sort of a greatly overgrown sea lion with a long sinuous neck.

Meanwhile, the hunt goes on with telescope camera and even biopsy darts, which if shot into Nessie would remove a bit of flesh for scientific analysis. Roy Mackal, the University of Chicago biologist who heads the American branch of the Loch Ness Phenomena Investigation Bureau, has even tried fishing for Nessie with a cannister of bait the size of a cookie jar, tethered by a ten-pound test line. Something bit the line through and made off with the bait.

Nessie is still with us after all these centuries, and so are its finny or furry relatives out in the salt water. Since the author has no intention of competing with the published lists boasting many hundreds of sightings, one modern sighting will close the case for the defense.

LAT 34.34 N LONG 74.07 W 1700 GMT STRUCK MARINE MONSTER EITHER KILLING OR BADLY WOUNDING IT PERIOD ESTIMATED LENGTH 45 FEET WITH EEL LIKE HEAD AND BODY APPROXIMATELY THREE FEET IN DIAMETER LAST SEEN THRASHING IN LARGE AREA OF BLOODY WATER AND FOAM SIGHTED BY WM. HUMPHREYS CHIEF OFFICER AND JOHN AXELSON THIRD OFFICER.

This wireless message was sent to the U.S. Hydrographic Department on December 30, 1947, by the Grace Liner *Santa Clara* sailing from New York to Cartagena, Colombia. The captain of the *Santa Clara* later submitted a more detailed communique, which made all the newspapers. Some zoologists said it was only an oarfish (it didn't fit the description at all), and the whole incident soon drifted into that oblivion where dwell all the unexplainable things seen by unacceptable observers from caveman to airplane pilot.

The official zoological opinion, if such exists, holds that sea serpents are cases of mistaken identity. Many big and rather strange animals are legitimate residents of the oceans. Logs and rafts of floating seaweed can look monsterlike under the right conditions. Dolphins, they say, can look like sinuous sea serpents when swimming in single file. But can we believe that men who have known the sea and its ways all their lives cannot recognize a school of dolphins? The evidence, though testimonial, is too strong, the witnesses too sincere and too consistent. It is more reasonable to believe that large animals unknown to science swim the seas.

Heuvelmans has sorted out 587 hard-core sightings into a series of bins based upon the reported characteristics of the suspected sea monsters. Over half of the sightings fall into such categories as mistaken identities, hoaxes, periscopes, and doubtful reports. There remain, however, more than 200 solid sightings divided into nine "species" of marine animals hitherto unknown to science. The most common (82 cases) is the so-called long-necked sea serpent, which is probably a mammal. Other probable mammals on the list include the many-humped sea serpents, merhorses, and the super-otters. The remaining six classifications represent large and, of course, very rare saurians, turtles, and super-eels.

Heuvelmans has been conservative in his analysis; this should always be emphasized. Yet, even with his high standards, he is able to present surprisingly thorough descriptions of giant, bizarre mammals, saurians, and fish unknown to science. Not only are descriptions consistent within each species, but geographical preferences are very marked for the super-otter, the many-finned sea serpent, and the many-humped sea serpent. The latter seems to opt for the Atlantic Coast of North America (remember Gloucester Bay) with additional sightings around Iceland and the British Isles. The ubiquitous long-necked sea serpent strongly resembles the Loch Ness monster. Perhaps a pair or two were trapped in Loch Ness in times past.

The long-necked sea serpent is seen most often. It is seen everywhere except the polar waters, but particularly around the British Isles and the North American coast between the end of April and

The long-necked sea serpent (really a mammal). Length: about 6o feet. (Reproduced with permission from B. Heuvelmans, *In the Wake of the Sea Serpents*)

October. Heuvelmans has noted the increased frequency of sightings and suggests that the long-necked sea serpent may be an ascending species. Obviously it is not a serpent but a mammal, probably related to the pinnipeds (seals and sea lions). The neck is long and swanlike, with a seal-like head and small eyes. It is furry, thickly covered with fat, and has flippers like the sea lions. The length varies between fifteen and sixty-five feet, and like others of its family the long-necked is a very fast swimmer. It probably hunts in the dark waters (like those of Loch Ness) with sonar, like other pinnipeds, and has scant need for good sight. It is the only sea-serpent species that has been seen on land.

A great deal more descriptive information has been collected by Heuvelmans for the long-necked and the other eight sea-serpent species. The evidence is almost overwhelming when taken in a single dose. Heuvelmans believes that testimonies from so many seemingly reliable witnesses familiar with the sea cannot be denied. On the concluding pages of his book *In the Wake of the Sea Serpents* is this paragraph:

> The legend of the Great Sea-Serpent, then, has arisen by degrees from chance sightings of a series of large sea-animals that are serpentiform in some respect. Some, like the oarfish, the whale-shark and Steller's sea cow, have been unmasked in the last few centuries. But most remain unknown to science, yet can be defined with some degree of exactitude, depending on the number and precision of descriptions that refer to them.

Sea serpents, once creatures of legend, are at last gaining some stature in the zoological world. Even after they are better described and collected for the museums and zoos, there will still be other mysteries beneath the sea—far, far beneath, as some of our sonar experience suggests. There have been several instances where several pieces of sonar equipment on different ships in a task force heard the sound of a great body propelling itself with some powerful mechanism at high speed through the black waters and crushing pressures five miles down. The speeds of these "things" seem to be as high as those of the most advanced nuclear submarines. The "thing" could hardly be an animal at these depths —or could it?

Instead of letting animal size and rarity be our guide, let us look at smaller creatures that pose monstrous problems. In this category are those "living fossils" (a term coined by Charles Darwin) that once disappeared from the strata that form the paleontologist's book of life and have lately been found hale and hearty. Take the coelacanth, the most famous "living fossil," a fish that was supposed to have become extinct some seventy million years ago. It is as alive today as the tuna and the mackerel, but it is rare, and you have to know just where to look for it.

Obviously many animals now walking the earth and swimming the seas are also part of the fossil record. Some reptiles and fish, for example, have been tough enough to survive whatever forces nature unleashed down through the ages. They do, however, form a continuous paleontological record. There are several implications of the coelacanth and the other living fossils that seem to have disappeared from recently deposited sediments:

The fossil record is far from complete. Actually, with more paleontologists in the field these days, many new discoveries and "missing links" are being uncovered. The coelacanth may actually show in the fossil record right up to the present, but we just haven't found the recent pages yet. For instance, the coelacanth may have been sheltered from evolutionary pressures in some restricted locality and thus have survived in small numbers despite the climate change or other calamity that killed the main population. The marsupials that survived in isolated Australia are good examples.

Another possible implication is that our evolutionary time scale is all wrong, as George McCready Price claims, pointing to his supposed "upside down strata." In his view there are no such things as *ancient* fossils; all life is essentially contemporary because it was all created in the same burst of energy by the Creator. That a few coelacanths could survive the Deluge would not be surprising. Unfortunately for Price, radioactive dating confirms that the earth's sediments are billions of years old. It is strange how so much in this book hinges upon one scientific technique. What if, as some cosmologists suggest, the clock of the universe ran much faster in ages past and is now slowing down and about to reverse? All the geological ages would then be compressed like an accordion instead of being stretched out for billions of years as our current geologic timekeepers indicate.

One other possibility remains: perhaps the coelacanth really did meet its end seventy million years ago and then, more recently, swam down the same evolutionary channel it did when it first evolved. Impossible, say the biologists, because evolutionary pressures would be different and life's germ plasm would have been diverted into another one of the near-infinite channels available to it. Besides there probably has not been enough time.

The first implication is by far the most reasonable one in the light of what we now know, although there may be some tiny bits of truth in the other two.

One aspect of "living fossils" that melds well with the history of sea serpents is the attitude of science in general to fishermen's reports of strange catches. Despite good descriptions, science had to have a coelacanth in hand before its existence could be admitted —the testimony of laymen could not be taken at face value.

The first important living fossil to be resuscitated, our coelacanth, has become a synonym for the acknowledged incompleteness of the fossil record. The first coelacanth to be caught in modern times (as far as we know) was pulled aboard a small fishing boat in a trawling net on December 22, 1938, about eighteen miles southwest of East London, a town on the southern coast of Africa. Among the three tons of fish hauled in during one pass was a five-foot steel-blue fish, with large dark blue eyes and—something rather unusual—a second dorsal fin. It was undeniably a strange fish, and the fishing boat captain turned it over to a Miss Courtenay-Latimer, the curator of the East London Museum. Unable to identify it, Miss Courtenay-Latimer contacted Professor J. L. B. Smith, South Africa's leading ichthyologist. Smith recognized the fish as an "extinct" coelacanth from the sketches that were sent him. It was unbelievable. Smith hurried to East London, but the flesh and organs of the coelacanth, which were decomposing in the summer heat, had already been disposed of at sea. Miss Courtenay-Latimer, however, had preserved the skin, the skull, and other parts with better keeping properties. She also made detailed notes on the fish's anatomy. In honor of her efforts, Professor Smith named the species Latimeria.

The coelacanth's skin and skeleton were important to science, but its soft parts, those organs that are hardly ever recognizable in fossils, would have been even more valuable to biologists trying to trace the evolution of the heart and similar organs. Professor Smith reasoned that the coelacanth fortuitously caught off East London was probably a stray from the deep Mozambique Channel between Madagascar and the African mainland. He therefore began a publicity campaign featuring leaflets showing a sketch of the coelacanth and offering rewards of one hundred pounds for each of the first two coelacanths turned in. At last, after fourteen patience-consuming years, on December 20, 1952, another coelacanth was caught by hand line off the island of Anjouan, in the Comoro Archipelago. Natives in the marketplace, where the coelacanth was about to be cut up for food, recognized it from the leaflets. When Professor Smith received the wire about the second coelacanth, the South African government provided a special plane for him, and he arrived at Anjouan just in time to preserve the fish for scientific study. Smith told reporters, "When I got to the island and knelt down to look at the coelacanth lying on its bed of cotton wool, I am not ashamed to say that I wept." Nearly a dozen more coelacanths have been caught since, but they are anticlimatic. The first two coelacanths—supposedly seventy million years dead—make a great story of scientific discovery. And we have hardly begun to explore the deeper waters of the oceans; or even the shallow waters close to home, for primitive fish scales the size of silver dollars turned up in a Florida souvenir shop one day recently. Some local fisherman, who could not be traced, had caught something much like the African coelacanth.

The coelacanth is merely a famous example of a phenomenon that recurs more and more frequently as we get to know our planet better. The continents in their meanderings have created refuges both above and below sea level where animals might escape the full brunt of the forces of evolution. Australia always figures strongly in talk about living fossils. When white men arrived, the whole continent was dominated by marsupials, except for the bushmen, the dingos, and some bats. In its rivers the lungfish, another missing link in evolution, was found. On the fringing shores, the "extinct" clam *Trigonia* was found; it had been abundant around Europe in the Age of Reptiles, but had long since been relegated to paleontological texts.

We tend to view these Australian animals as lucky survivors because they are "primitive," that is, they have not been "forced to evolve" into the modern forms found on the other continents. Yet the marsupials and their primitive colleagues prospered as Australia drifted away from mainland Asia; in fact, in terms of longevity, they are more successful than modern mammals, being admirably adapted to Australia's environment. Evolution never tears down its old structures completely, particularly if there is no need to. It is wise to maintain many lines of defense, many diverse gene pools, against whatever environmental catastrophes may arise.

READING LIST

BURTON, M.: *Living Fossils*, Vanguard Press, Inc., New York, 1954.

CARRINGTON, R.: *Mermaids and Mastodons*, Rinehart & Company, Inc., New York, 1957.

DINSDALE, T.: *Loch Ness Monster*, Routledge & Kegan Paul, Ltd., London, 1961.

—— *The Leviathans*, Routledge & Kegan Paul, Ltd., London, 1966.

GOULD, R.: *The Case for the Sea Serpent*, Phillip Allan & Co., London, 1930.

HEUVELMANS, B.: *On the Track of Unknown Animals*, Hill & Wang, Inc., New York, 1959.

—— *In the Wake of the Sea Serpents*, Hill & Wang, Inc., New York, 1968.

HOLLIDAY, F. W.: *The Great Orm of Loch Ness*, W. W. Norton & Company, Inc., New York, 1969.

LEY, W.: *The Lungfish, the Dodo, and the Unicorn*, The Viking Press, Inc., New York, 1948.

—— *Exotic Zoology*, The Viking Press, Inc., New York, 1959.

OUDEMANS, A. C.: *The Great Sea-Serpent*, Luzac & Co., London, 1892.

THOMPSON, C. J. S.: *The Mystery and Lore of Monsters*, University Books, Inc., New Hyde Park, N.Y., 1968.

I I

RECAPITULATION

Exploration of the sea and the sea floor has forced scientists to reconsider the role of Catastrophism during the evolution of the earth, especially the last several hundred million years. Continental drift, sea-floor spreading, guyot formation, and most of the other phenomena discussed in the preceding chapters may eventually be brought into the Uniformitarian fold by enlarging the definition of Uniformitarianism. After all, the only physically measurable difference between Uniformitarianism and Catastrophism is the time during which the forces of change act. (The philosophical differences will probably never be eliminated.) On the other hand the tektites, the huge meteor craters, magnetic reversals, several astronomical phenomena, and perhaps biological extinctions seem more Catastrophic than Uniformitarian. And it may be that some essentially Uniformitarian processes, such as sea-floor spreading, were initiated by Catastrophism. In any case, the earth seems to be a much more plastic planet than hitherto supposed.

Just how far the principle of Uniformitarianism would have to be expanded to encompass the new evidence from beneath the sea and from associated astronomical phenomena is evident from the following table. Most of the evidence for frequent, widespread Catastrophism must be classified as *weak*. Donnelly, Price, and Velikovsky are not vindicated at all by the facts we know today. This same evidence, though inconclusive as most of it is, could never have been accepted by a Uniformitarian in 1950. Earth scientists

156

and astronomers are breaking the grip of doctrinaire Uniformitarianism. The liberally minded Uniformitarians, who admit that Catastrophism has played a significant role in the earth's evolution, are nudging closer and closer to the objective Catastrophists, who realize that the past has not been all fire and brimstone. With new facts being dredged continually from the sea bottom we may eventually see a ceasefire in this two-centuries-old conflict.

POTENTIAL CAUSES OF CATASTROPHISM	SOLAR SYSTEM EVIDENCE	MAGNETIC EVIDENCE	DRIFT AND SEA-FLOOR SPREADING	TERRESTRIAL METEOR CRATERS
Celestial encounters (meteors, comets, planetoids)	(S) Lunar and Martian craters. Resonance of Venus	(S) Tektite ages correspond to field reversals	(W) Birth of moon or meteor impact *might* start drifting	(S) Several large eroded craters known; viz., Hudson Bay
Solar storms (plasma, cosmic rays, light)	(S) Fused areas on lunar surface Frequent flares	(W) Plasma might cause field reversal, laboratory evidence		
Supernovas		(W) Ditto		
Terrestrial radioactive heat		(W) Magnetism affected by heat in laboratory	(W) Heating, expanding earth a *possible* interpretation	
Internal mass shifts within earth			(W) *Inference* only, no evidence	
Self-reversal of earth's magnetic field		(W) Self-field reversals noticed in laboratory for some rocks		
Extensive volcanism				
Man-made pollution and environmental changes				
Pandemics				

W = weak case for Catastrophism
S = strong case for Catastrophism

TEKTITE FIELDS	ICE-AGE EVIDENCE	SEA-LEVEL CHANGES AND WIDESPREAD FLOODS (GUYOTS, SUBMARINE CANYONS)	EXTINCTIONS OF LIFE (FOSSIL RECORD)	LEGENDS, MYTHS
(S) Several tektite fields indicate catastrophic origin	(W) Encounter *may* have changed pole of rotation	(W) Flooding *might* have been caused by an encounter	(W) Orogeny, floods, etc., implied	(W) Many, many tales of celestially stimulated havoc
	(W) Drop in solar constant *might* have caused Ice Ages		(W) Radiation extinction possible	(W) Ditto
			(W) Ditto	(W) Ditto
	(W) Change of pole of rotation a *possible* cause	(W) Changes of angular momentum *might* cause flooding	(W) Orogeny, floods, etc., *implied*	(W) Many, many tales of terrestrial havoc
			(W) Radiation extinction *possible* when field disappears	
	(W) Volcanism *might* have reduced heat reaching surface		(W) Accompanying orogeny *might* have caused some extinction	(W) Ditto
			(W) A few recent extinctions noted	
			(W) No direct evidence	

INDEX